REMOTE SENSING TECHNIQUES FOR REGIONAL DEVELOPMENT

About the Authors

Dr. Ranjit Kumar Banerjee, M.Sc. Ph.D. is one of the senior geographers of India with specialisation in Thematic Cartography and Mapping. Born at Narayanpur, Barisal (now in Bangladesh) in 1935, Dr. Banerjee had his school education there. He was graduated from the University of Guwahati and obtained his Master's degree, as well as his Ph.D. degree in Geography from the University of Calcutta. Immediately after he joined The National Atlas and Thematic Mapping Organisation, (NATMO), Government of India, and retired from that organisation as Joint Director in 1993.

He is one of the pioneer geographers nurturing the growth and development of NATMO from its infancy to its present state. During the period of his services, Dr. Banerjee was deputed to take specialised courses on Photo-Interpretation in Soil Survey from IPI, Dehradun and obtained first class diploma from ITC Delft, Netherlands and also obtained a first class certificate in Digital and Analogue anlysis and application of Remote Sensing in the field of agriculture, forestry, hydrology etc. at aerospace centre, FAO/DSE/DFVLR, Germany. He also participated in a number of national and international conferences such as in Sri Lanka, Brazil, China and West Germany. He was visiting lecturer in Jadavpur University, 1981, Calcutta.

Dr. Banerjee is also associated with a number of pioneer educational research institutions of this country. He is presently acclaimed more as a Remote Sensing scientist rather than a Geographer. He published more than 10 research paper in National and International scientific journals. Dr. Banerjee is also life member of INCA and Geographical Society of India. He has extended immense contribution towards thematic Cartography by participating and directing a number of Cartographic projects during his tenure in NATMO.

Professor Bireswar Banerjee is a renowned geographer of international celebrity. He is the first to obtain D. Lit. Degree (University of Calcutta) from India. He is well versed in various branches of Geographical sciences particularly in the field of Social and Cultural Geography, Environmental Geography, Geography of Social Well-being, Economic Geography, Regional Development and Planning. He is associated with the University of Calcutta, Geograpy Department as Lecturer (1963-72), Reader (1973-77) and Professor (1977-95) and Head of the Department (1983-85) and presently as UGC Project Coordinator (as retired Professor).

He has supervised a number of research projects both independently and in collaboration with international institutions. He has already guided the works of 35 research scholars and more are in the line. He is associated with the academic programmes of various Universities of the country as well as with various national organisations like UGC, INSA, ICSSR and others. He has more than 100 research papers published in National and International Scientific journals. He was visiting Professor at the University of Hiroshima in Japan in 1993 and at the University of Nagoya, (Japan) 1999. He is still engaged in various research and academic activities.

REMOTE SENSING TECHNIQUES FOR REGIONAL DEVELOPMENT

R. K. BANERJEE
Joint Director (Retd.)
National Atlas and
Thematic Mapping Organisation
Calcutta

BIRESWAR BANERJEE
Professor of Geography (Retd.)
Department of Geography
Calcutta University
Calcutta

CONCEPT PUBLISHING COMPANY, NEW DELHI-110 059

ISBN 81-7022-838-7 (PB)

First Published 2000

Published and Printed by
Ashok Kumar Mittal
Concept Publishing Company
A/15-16, Commercial Block, Mohan Garden,
NEW DELHI-110059

Phone : 5648039, 5649024
Fax : 091-11-5648053
E-mail : publishing@conceptpub.com

DEDICATED
TO
LATE PROF. MAYA BANERJEE

Head of the Department,
Lady Brabourne College, Calcutta.
Govt. of West Bengal

PREFACE

There are various landforms in our country from lofty mountains of the Himalayas, vast plain lands, large coastal areas, but we cannot access their potentialities as a resource base and use them efficiently. We get good rainfall every year but we cannot store them properly. So dams are being constructed at vantage points for agriculture, hydro-electricity etc.

The aim of this study is to portray how Aerial Photographs, Satellite Imageries and other Sensors could properly access the potentialities of the land. To illustrate such an assessment effectively three regions of mountains and hills, plateaus and coastal areas have been chosen. This book gives a comprehensive understanding of the complex interaction between the man and environment. People of Orissa realised at a later stage that the irrigation canals particularly by Bargarh and Sason canals of Hirakud dams unfolded tremendous potentialities to boost the economy of Orissa particularly as rice producing areas. Government of Orissa, particularly the Administration of Burla helped me tremendously to the success of the book.

The case study of Mahanadi Delta gives a vivid account of potentialities of the land and the sustained production of rice, a resultant of the Hirakud Project. 192 MW of hydro-electricity is produced to boost the infrastructural and ancillary industries of the region. The scenario has been completely changed today. The Government of Orissa, provide data supplementary to the interpretation of Aerial Photographs and Satellite Imageries. The areas, which have been studied, have scanty information and very limited research work has been done in these areas. In this book, a combined effort is made to bring out the reality of the picture and very clear understanding as a whole.

I am grateful to the U.P. Government for providing me PWD rest-house even in jungles due to which the study of Hardwar and Rishikesh area had been possible. Some of these are the Motichur Forest Duk Bunglow and Forest Duk Bunglows of Forest Department. I express my due gratitude to the Survey of India for providing Departmental sub-staffs for camping in different localities. I also express my deep gratitude to the Indian Photo Interpretation Institute (IPI), Dehradun, for providing consultations and guidance by the Dutch experts (A.K. Kamphorst, Dr. Nossin); and also for providing me vehicles during the survey work.

The work reveals the Chandna Rao terrace 4, which is the product of predominant fluvial action in the past by Ganga and its tributeries such as Chandna Rao, Song and Jhakhan Rao. The results of analysis reflect clear geomorphologic units, and the Satellite Imageries (Band 5, Band 7), and Geo-coded Imageries on a larger scale show the clear division between the Lesser, Middle and Greater Himalayas. The different soil types of the separate units have been analysed through interpretative methods. Lastly a summary and conclusion has been drawn that the Aerial Photo Survey, Satellite Imagery Survey and Collateral Information had unfolded the whole evolutionary history of the area. During the entire fieldwork, we were helped tremendously by the PWD engineers, Forest Range Officers and Military Personnel.

Authorising this book would not be possible if not for a tremendous encouragement by

Prof. Bireswar Banerjee of the University of Calcutta. He helped me in every step, for rectification and beautification. I am really indebted to him for his help. Secondly, I am deeply indebted to my wife, late Prof. Maya Banerjee, Head of the Department of Geography, Lady Brabourne College, Calcutta, for her outstanding help, constant inspiration and encouragement as well as invaluable contribution towards my thesis, specially involving the mathematical calculations and overall effort. My dream has been fulfilled due to her constant guidance and immense sacrifice.

I also express my gratitutde to the authorities of the NRSA and Survey of India for permitting me to consult Satellite Imagery and aerial photographs. The help and encouragement rendered by Dr. B.K. Roy, Director, National Atlas and Thematic Mapping Organisation is gratefully acknowledged.

I express my deep appreciation to Sri Manas Kumar Maity, Scientific Officer NATMO, who helped me in the fieldwork in Sambalpur District. I also express my deep gratitude to Sri Sunil Kumar Ghosh, Scientific Officer and Sri Sujit Kumar Banerjee, Senior Research Assistant, NATMO for helping me in the preliminary stage of the field survey of the same area. I express my thanks to Sri R. Das, Chief Engineer of the Hirakud Dam Circle.

The cartographic assistance and typing rendered by Sri A. K. Mitra, S.C. Das, S. Chanda, A.N. Dey, George Mariano, M. Talukdar and B. K. Roy Chowdhury are hereby acknowledged. I also express my thanks to Dr. Satyam Kumar Kundu for his meticulous final typing of this thesis and bringing it to final shape.

My daughter, Dr. Anuradha Banerjee, Assistant Professor in the Centre for the Study of Regional Development, School of Social Sciences, Jawaharlal Nehru University, New Delhi; who helped me in all possible ways in the orientation and editing of the book. My son Amitava Banerjee had assisted me constantly, by appreciating my gesture of undertaking such a work.

As earlier mentioned my wife Late Prof. Maya Banerjee had extended her constant flavour and love for the subject by upgrading the book to its present standard. Without her inspiration I would not have been successful and thereby I dedicate the volume to her loving memory.

Date: 09-10-1999

Dr. RANJIT KUMAR BANERJEE

Calcutta

CONTENTS

LIST OF MAPS AND DIAGRAMS

LIST OF TABLES

LIST OF PLATES

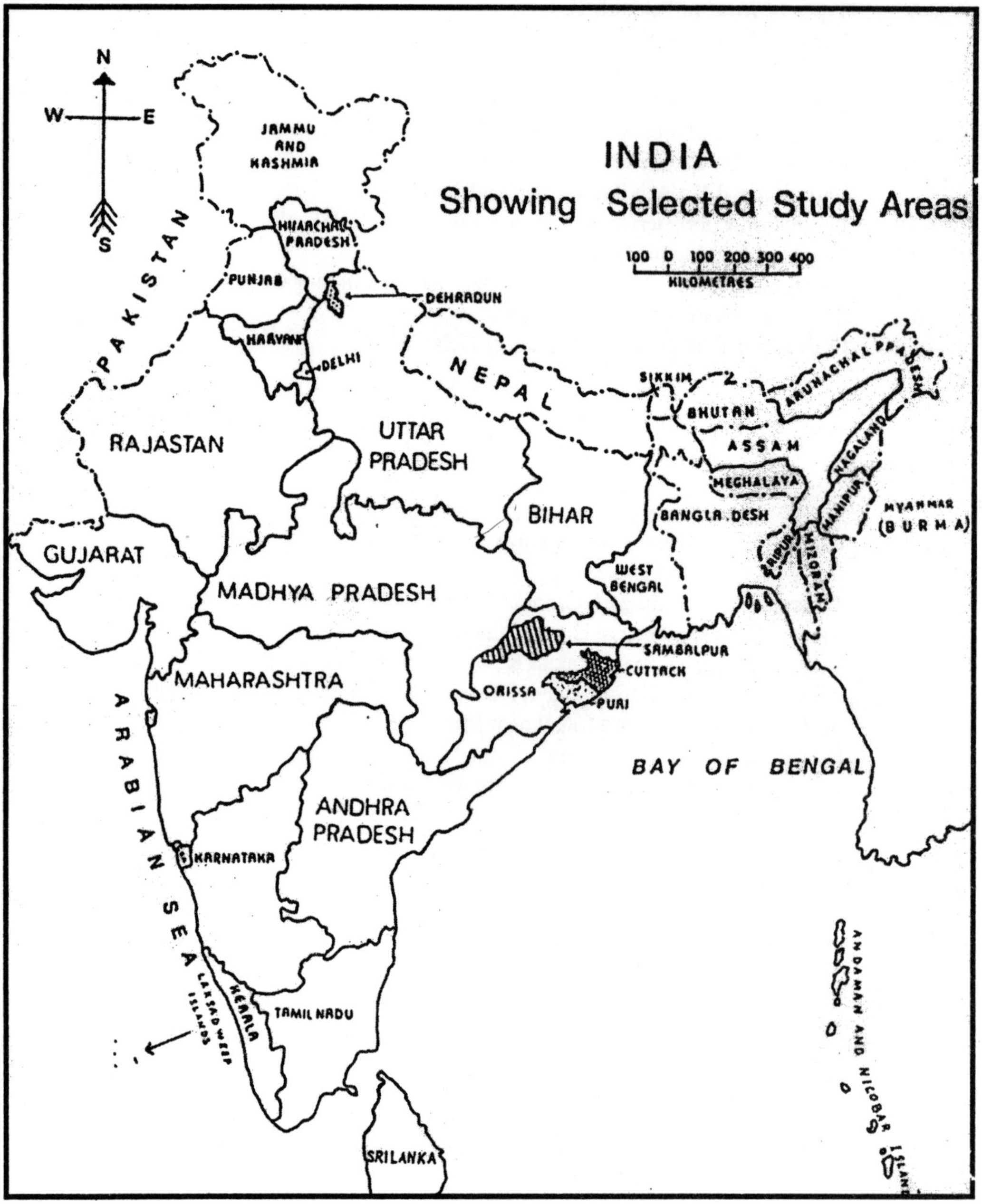

Location Map

1

INTRODUCTION

Any type of study of earth such as survey, analysis, terrain, study have now become more easy by recent development and principles of Remote Sensing Technique. Remote Sensing is the acquisition of information about an object without physical contact. The word remote sensing was coined by geographers during 1960 in the Office of Naval Research, USA. Earlier any type of survey is made in conventional way; that, one has to go personally to the field and to do the work on the basis of information collected. But with the availability of aerial photographs and Satellite Imagery the geographical studies have become easier. In case of Remote Sensing Technique one can get information of inaccessible area and maps can also be drawn in the laboratory. Visual images of large areas enables the observer to perceive relation between objects and their environmental surroundings which to a ground observer, might not be evident. The photo and imagery interpreter can study significant pattern without the distraction of irrelevant details and can compare these patterns with map representation. It is also important that every detail or element can be studied clearly under different instruments like pocket stereoscope, mirror stereoscope, Multispectral Digital Analysis System (MDAS) etc. Aerial photographs are used since 1914 and from 1972 Landsat Imagery gives an additional tool. Landsat Imagery is now considered a powerful tool in geographical research as it provides more synoptic coverage giving better knowledge of larger areas. From aerial photographs and Satellite Imagery, a new science has been developed known as 'Interpretation Science.' This science is recently developed and from this details of terrains can be worked out.

Aerial photo interpretation science and other remote sensing techniques followed by USA, Russia, India and others are now being advantageously used in interpretating the hydrology, landuse characteristics of inaccessible as well as accessible terrain. The water retention capacity of the soil can be deciphered with the help of aerial photographs. These tools are also useful in preliminary ingestigation for alignment of irrigation channels, power houses, dam sites to a reasonable degree of accuracy. The terrain evolution and delineating pattern could be extended by extrapolating without doing the conventional experiments in the field.

Evolution of the Concept of Remote Sensing

The real meaning of the photogrammetry is measurement on the photographs, imagery etc. Gradually Interpretation Science is very rigidly transferred into Remote Sensing. The meaning of Remote Sensing is already defined at the begining of this chapter, it is again clarified here to undergo the indepth study to understand the concept of the subject. The actual definition of Remote Sensing is the observation of a target separated from it by some distance, without physical contact. The actual importance of the physical link between the objects of measurement and remote sensing measuring devices involve electromagnetic energy, acoustic waves, and force fields associated with gravity and magnetism. Both acquisition of information and measurement are involved in Remote Sensing Technique. Further it may be well said that, if the remote sensing is being allied with cartography through the information system, Remote Sensing Technique can rise above the level of a mere technology. The actual concept refers to the gathering and processing of information about the earth, its environment, particularly its natural and cultural resources by various modern instrumental techniques. The objective of this chapter is to provide an overview and historical perspective of remote sensing and to introduce the more important basic concepts, sensors and its application, data processing, analysis and utilization methodologies for the purpose of research. Finally, it has been conceived that it now provides exciting windows as multidimensional concept. So with the discovery of new instruments Remote Sensing Technique occupies an important place in the space science. So it would be wise to examine the development of remote sensing. It is observed that most of the remote sensing instruments are developed by developed countries like USA, Canada, East Germany and West Germany, Switzerland and Japan. But above all the countries, the Carl Zeiss Jena of Germany became pioneer in this field. The new factory of Carl Zeiss was set up at in Overkochen in West Germany and the production of the West Germany has been found to be more sophisticated than the original Carl Zeiss Zena's production, particularly stereoscope, stereopantometer, sketch master, slotted template and many others. Now these instruments are developed in the developed countries and the developing countries are in completely dark about those. So transfer of technology is the only alternative. Many institutions have come up in the developing countries with their Remote Sensing equipments. The Dutch extended their collaborations in setting up Indian Photo-Interpretation Institute (IPI) (now Indian Remote Sensing Institute). Subsequently, the Institute came under the Department of Space, Govt. of India. Similar institutes have also been set up in the Far East viz., Indonesia (LAPAN) and Asian Institute of Technology at Bangkok.

In the North of India lies the Himalayas. Recently, Wadia Institute of Himalayan Geology has been set up for indepth study of the Himalaya Scientists expect good amount of treasure underlying the rock strata. The valuable works of geologists like Burrard and Hayden, Gansser, Everest, Oldham, West, Wadia and others have contributed much to, unveil the structure and relief of the Himalayas. But in recent years with the advent of aerial photographs, space imageries, shuttle flight photographs, enormous data are gradually been utilized by foreign scientists and they prepared very good maps of Nepal Himalaya and the Himalaya as whole, in this respect German scientists forged ahead with the activities of the full utilization. What wonder the German scientists produced in The German Aerospace Research Establishment (DFVLR) in Oberffafenoven by shuttle flight photographs which has the effect of stereoscopic vision and Mount Everest has been pin pointed in the stereopair and the idea of such magnanimity is really understood. Similarly, the inaccessible parts of the Himalayas are now being explored by Remote Sensing Technology. But Wadia Institute of Himalayan Geology is doing good work, proper synchronisation is needed for further indepth studies because to unveil the structure, relief, drainage and other elements of this formidable mountain system.

Aerial photographs and Satellite Imageries are now being used in India in many departments. But their application should be carefully made, it may be noted that such tool is not required everywhere for the same purpose because the respective area has already been studied by the conventional surveys. We have used the tool in many places. At places Remote Sensing Technique is found to be superior than others. For example in the Himalayan and Sub-Himalayan areas, the technique is found to be very useful particularly in geological investigation.

Selected Areas

As such different areas have been selected in this study. The basic objective is to portray the effectiveness of Remote Sensing in area studied. The areas selected are :

(1) Rishikesh Police Station area near the Himalayan foothills.
(2) Hirakud Canal Command area, under the Sambalpur District of Orissa State.
(3) Mahanadi River Delta covering partly the area of Cuttack and Puri Districts of Orissa.

Himalayan Foothills

Limited area of the Siwaliks foothills near Rishikesh Police Station of Dehra Dun district is more or less bounded to the South by 350m contour line. This region is drained by three important streams. This foot hill zone is adorned by vast expanse of alluvial fans. Another characteristic feature is the occurrence of terrace formed by the sides of the Chandna Rao and the Ganga river. Such alluvial fans and terraces are clearly deciphered from air-photographs and Satellite Imagery. The Song, Jakhan Rao and its tributaries play important role in shapping the physiography of the area. On the basis of information thus gathered, photo-interpretation of the region is attempted and subsequently a detailed soil map has been also prepared. This is also supplemented by field observation.

It has been ascertained that air-photographs save considerable time because the interpretation can be carried out in laboratory. To make the map more authentic, field survey is carried out covering 10 per cent of the region under investigation. From the economic point of view, 80 per cent of conventional field survey could be reduced and 80 per cent of time could be saved. Similarly, in case of Satellite Imagery 10 per cent of each scene are studied in detail by "Supervised classification" and further by extrapolation by "Unsupervised classification." The supervised calssification means a computer implemented process through which each measurement vector assigned to a class according to a specified decision rule where the possible classes have been defined on the basis of representation training samples of known identity. These known identities are extrapolated thereby enabling the researcher to reduce the time of field work by 80 per cent and the cost of survey by 80 per cent as well.

Hirakud Canal Command Area

Situated in the Mahanadi Basin the area selected is the second micro-region for comparative analysis. Unlike the Himalayan foothills, there are little variations in relief. The Mahanadi river valley is noted for extensive landuse system. The region contains most fertile farmlands, producing fairly good amount of rice. So Hirakud project and its Canal Command Area irrigates 2.67, 000 ha. of land. The study of physiography of this area is not very useful as the land is almost flat with a few residual hillocks which are mostly accessible. The socio-economic set up of the Canal Command Area has been much improved due to the canals of the Hirakud, and its distributaries. The lower reaches of Puri and Cuttack districts are saved from devastating floods of the Mahanadi, the Baitarani and Brahmani as all rivers are controlled by Hirakud and other dams. From this area, 3 towns and 6 villages have been picked up for detailed analysis. The towns are Bargarh, Hirakud and Burla and the six villages are (i) Jamadarpali, (ii) Balbaspur, (iii) Gurbagha, (iv) Ladarpali, (v) Kurla, (vi) Suktapali. In the Canal Command Area, three crops mostly paddy are cultivated. These are then exported to the neighbouring states. Modern methods of farming are practised here. It has been also noticed that by the side of big canal, like Bargarh and Sason canal, green belts have come up. These are vegetable farms. Green vegetables like pumpkin, papaya, palbal etc. are exported to Cuttack, Calcutta and other places by trucks. It is also been observed that after construction of Hirakud Dam, two new towns, viz., Burla and Hirakud have come into existence.

Burla develops as administrative town where Medical College, Engineering College and Sambalpur University are located. Hirakud Dam Project Office is situated at Burla which controls the function of this Dam. The Hirakud reservoir one of the largest reservoirs of India. So with the help of Remote Sensing Technique the area has been studied. Further, the latest Landsat Imagery shows that due to construction of several dams over the Mahanadi and the Brahmani, the siltation of the Mahanadi Delta has been considerably checked. In certain areas the saline encrustation on land is increasing. This is clear from the Landsat Imagery. Hirakud town is being gradually developed as an industrial town.

It has thus become clear that two regions selected represent two distinct regions of diverse characters. The Sub-Himalayan region is noted for its distinct land form elements. These elements can be better understood with Remote Sensing Technique. The Mahanadi on the other hand is characterised by diverse landuse, the intricacies of this diverse use can be accurately deciphered with Remote Sensing Technique. But conventional survey has also been carried out in the 3 towns and six villages within the specific time.

The Mahanadi River Delta

The third selected area chosen is near the coast. The area covers the deltaic portion of Cuttack and Puri Districts of Orissa. The Mahanadi in combination with the river Brahmani and the river Baitarani has been able to form an extensive alluvial tract extending from Chilka lake in the south to Balasore in the north, 175km long and 75km wide. A distinctive feature of the growth of delta is that it has its maximum growth not in the centre but in the northern part. Prior to the formation of the delta the original shore line was concave towards the sea owing to the hard bed rocks of the Eastern Ghats. So these things are clearly revealed in the MSS band and sketch map of landform is thus prepared.

Advantage of Remote Sensing Technique

There are certain advantages also of using Remote Sensing Technology over conventional survey. The use of air-photographs for soil survey is in fact comparable to the use which any soil surveyor makes of the correlative indication in the field. The air-photo however has in this respect some very definite advantage.

(1) First of all it is possible using air-photos to analyse systematically the whole area to be surveyed.
(2) Secondly without going to a hill, rugged terrain or in a dense jungle the researcher can prepare a preliminary map with sufficient accuracy.
(3) The whole area to be covered can be viewed at a time and plan could be made accordingly.
(4) The possibility of planning the field work can be done in the area. It can be chalked out in advance taking care of the indication of the physiography that has been studied with the help of air-photos and Imagery.

Technique of Interpretation of Satellite Imagery and Aerial Photographs

The basic principles of interpretation are based on observation. The technique of observation is a very difficult task. One must apply logical modes of thought to draw correct conclusion from the things observed. To have a realistic picture, Satellite Imageries and vertical air-photographs are used. The following characteristics are given attention for interpretation of photographic images (1) Size, (2) Shape, (3) Shadow , (4) Tone, (5) Colour, (6) Texture, and (7) Pattern.

For interpretation many complex and costly instruments are used for special purpose interpretation. Firstly, three items of equipments are essential, a glass mark coloured pencil, a mirror stereoscope and a pocket stereoscope (particularly for field) and measuring devices. The function of the stereoscope is for visual requirement as it presents a fine model of the terrain. The lenses should be of good quality with no objectionable distortion characteristics and of high resolving power. A China pencil or grease pencil is often convenient for drawing on the photographs. It can be used on glossy as well as on semi-matt finish and the mark can be removed by rubbing with a soft cotton. The basic measuring device is a simple plastic or box wood rule. For more specialized analysis there are devices which help the interpreter to measure vertical heights.

Handling of Air-photographs and Imageries

A photo interpreter may have thousands of air-photographs on his desk at one time. This mass of air-photographs should be handled in an orderly way. The haphazard way of examining photographs is known as Fishing Expedition. Since the photographs are numbered, orderly arrangement is easy. Flight Engineers taking photographs in a planned way, which is called Task No/Specification No., Run No., Photo Number, etc., Thus, this facilitates of handling photographs by the experienced interpreters. They find out required photographs and make overlay and required photographs are interpreted methodically. Stereoscopic vision through pocket and mirror stereoscope is exceptionally important to the photo-interpreters. A person who lacks stereoscopic vision has to judge depth by means of monocular clues.

The following are the methods of interpretation with the help of stereoscope :

(a) Mark with a dot or pin the principal point (Geometric centre) of each photograph. This point is located at the intersection of perpendicular lines between the fiducial marks at opposite edges of the photographs;

(b) Mark on each photograph the principal points of the other photographs of the stereoscopic pair. On both photographs draw a straight line between principal point and conjugate principal point. This line is a segment of the flight line. The two exposures are made from camera stations directly above the two principal points;

(c) The photographs are overlapped so that the corresponding images are superimposed and extend from left to right with respect to the observer; and

(d) Place the stereoscope over the photographs so that the lenses are aligned with the flight line and above the two images to be viewed. If a mirror stereoscope is used, the photographs are separated by a distance equal to that between the centres of the wing mirrors. Usually, one adjustment will not serve for examination of all parts of the stereoscopic model. Particularly in photographs having a short principal distance, radial displacement requires readjustment when the interpreter wishes to view images near the edge of the model. One of the photographs may have to be moved in the direction parallel or perpendicular to the flight line. The two flight line segments should be kept parallel. The lenses or mirrors of the steoreoscope should be directly above the images to be viewed. With experience, the need for readjustment is signalled by a sensation of stain or pull on the eye. The skilled interpreter can orient stereoscopic pair without recourse to the finger method. He could fuse the images perfectly by his experience and scan the whole area to be studied can mark the desired objectives with glass mark pencil and can draw the maps on the one of the photographs, right or left. Regarding the height measurement the interpreter use the instrument parallax bar and by computation of formula, flight determination could be made.

Methods of Search

A job of interpretation begins with a close examination of all details which are relevant but most experienced person prefers to begin by scanning the area as a whole or a large part of it. They are then prepared for intelligent selection and study of details. There are two ways to study aerial photographs.

(1) Fishing Expedition
(2) Logical Search

At first the interpreter is tempted for Fishing Expedition and thereby he knows the entire gamut of both, physical and cultural elements. As the basic principle is one topic to be studied at a time, thereby the scientist prepares a series of maps, such as (1) physiography, (2) vegetation, (3) Arable land etc. Further by collateral information the desired maps such as soil maps, geomorphological maps, geological maps etc. can be drawn. The collateral information are of three kinds (1) topographical sheets, (2) field observation in accessible area, and (3) literature of the area. The collateral information are of tremendous importance for a photo-interpreter. As a whole the photographs/ imageries speak of the world such as tropic, humid tropic, desert etc. Thus an interpreter brings the convergence of evidence and further he could prepare photo-interpretation key. A key may be organized for identifying by selection or by elimination. By this way of identification and the photo-interpretation, beautiful maps can be drawn from air photographs.

Interpretation of Satellite Data

The most important thing that Satellite Imagery gives is the repetitive coverage of a particular place in every 18 days. The orbital coverage creeps west-ward by one orbit swath each day, providing adjacent coverage on consecutive days. In some cases repetitive coverage varies from 14-18 days. LANDSAT 4, July 16, 1982 which is accordingly World Reference System (WRS) was injected into its desired orbit on July 29, 1982. This WRS orbit has a 16 day repetitive cycle, 233 orbits per cycle. So every 14 days the interpreter gets the photographs of the same place and possible changes are detected and could be mapped. The visual interpretation is more or less the same as in the aerial photographs, but for detailed study computer microprocess or with the computer compatiable tape helps for further study, but such facility is not available every where. It also requires detail training. As most of the Landsat Satellites and other Satellites move at 900km above the surface of the earth and the resolution varies from 80 to 8 meters. Only important thing is that it gives good synoptic coverage. But many processes are involved in Remote Sensing Technology by which Analog pictures could be converted into Digitized mode by computer and *vice versa.* But it is easy to obtain an Imagery of a particular place by identifying the number of path and row. But most of the work of the Remote Sensing Technique of Satellite Imagery are being done by MDAS/MSS of the ground station. Floppies are available for the area if the organisation do have the system.

Previous Works

Serious works on aerial photo-interpretation in India was initiated in India during 1965 after the establishment of Indian Photo-Interpretation Institute (IPI) at Dehra Dun at the instance of Survey of India with the collaboration of International Training Centre for Aerial Survey (ITC) Netherlands. During the 1960s the interpretation of such elements as Geology, Forestry and Soils were efficiently handled by Dutch Experts like Dr. Nossin, Dr. Kamphorst, Dr. Roupke and others. Subsequently photogeography has been adapted as an aid in preparing maps and other papers. Series of experiments were made with the aerial photographs. But in 1974 after the availability of Satellite Imagery from NASA, USA, NRSA Hyderabad, new vista has been opened because it is one of the strongest tool of research for the field Geographers. Isolated work on this topic have been done by various geographers. Details of this have been compiled in the Manual of Photographic Interpretation published by American Society of Photogrammetry (1960). In India lots of work on Photo-geography and Photo-geology are going on particularly in National' Atlas and Thematic Mapping Organisation, Geological Survey of India, Agricultural Research Institute (IARI) , All India Soil and Landuse Survey etc .

Photo-geography in the United States has made good progress since war. Many colleges and universities offer courses in photo-interpretation to the students of Geography and other Earth Sciences. Photo-Geographers also play significant role as compared to the Photo-geologist or other sister sciences.

Scope of the Work

The photo-geographers usually rely on air photographs and satellite Imageries in tackling geographical analyses. In this dissertation the researcher has based his observation solely on air photographs and some collateral information have however been incorporated here to magnify the subject. According to American Geographers there are three principal reasons of their reliance on aerial photography and Satellite Imagery than their colleagues in other sciences. The three reasons are :

(1) There are some geographers, whose training predates the development of photo-geographic methodology, they are unacquainted , with the extent of aerial photographic coverage, how to obtain photography of regions of interest, or how to interprete the photograph once obtained. This type generally belongs those associated with Indian Universities;

(2) The photo-geographers work is unusually complex. He may be well acquainted with all images relating to his own field of interest. But since geography deals with all phenomena on the earth's surface theoretically he should also be able to identify and interpret almost every image on any photograph. He must be selective in choosing the images for his requirement. The object of this study at first is confined to study the landform along with underground geology. In this particular field photo-geology of Dehra Dun area has been studied in detail. Second study is confined to and use of the Hirakud Canal Command Area where six villages and three towns have been analysed in detail with the help of aerial photographs and Satellite Imagery of Band 4567(MSS) on False colour composite;

(3) The third is the most important. The work of geographers is more often institutionally sponsored than commercially demanded.

So my study with the help of aerial photography is confined to three regions (1) One in the foot hills of Siwaliks in Rishikesh Police Station; (2) Hirakud Canal Command Area, and (3) The lower course of Mahanadi in deltaic area with the help of MSS band. They are detailed below :

(i) Photo-geographical study of Rishikesh Police Station which is a part of Dehra Dun district with the intention of preparing a soil map with photo-geographic technique;
(ii) To study landuse of the Mahanadi River Basin particularly of the Hirakud Canal Command area consisting of six villages and three towns with the help of aerial photographs technique and collateral information; and
(iii) To study the lower course of the Mahanadi particularly in the deltaic area with the help of Remote Sensing Technique.

After step by step of critical studies, viz., recognition and identification, analysis, deduction, classification of the desired subject in either landform or landuse, structure, relief etc. and after full establishment of their identity, we bring out conviction and prepare maps and description of the area through research.

The main thrust of this dissertation is concept of "land classification." The geographical definition of ' land' is everything on the surface of the earth. Photo-geography made the task easier to unveil the land surface, but from agricultural point of view, soil is seen from the stand point of the former. Geologist interprets the land from different point of view and their only idea is to find out indepth structure and relief. But it has been observed that they attach more importance to dissection and analysis of the drainage system. Because the drainage system on the sandstone is completely different from slaty surface and in the limestone topography missing drainage is prominent.

The basic analysis of this dissertation reveals a clear picture of the scope of Remote Sensing in geographical studies be its geomorphological, economic, landuse or integrated area analysis.

Though Aerial Photographs and Satellite Imagery are the main tools of the study, but collateral information and field works are made to assess the correct analysis of the selected areas. Maps are prepared as per desire with the help of Air Photographs and Satellite Imagery. So selective elements are chosen as per photo-interpretation methods. Few factors are considered carefully such as layout designs and scale factors etc. to complete the study as most comprehensive.

2

GROWTH AND EVOLUTION OF REMOTE SENSING

Elaboration of the Concept

The main objective of this chapter is to provide an overview and historical perspective of Remote Sensing and also to portray the basic concepts about the application of Satellite Imagery and photo-interpretation with respect to geographical analysis.

The word photo-geography is applied by J.H. Roscoe in 1952 and he delivered three papers related to photo-geography and as such he did not use the term "Photo-geographical analysis." *The actual definition of photo-geography is the joint efforts of geographical competence and photo-interpretation and its analysis keeping geography in the perspective.* So far the aerial photographs are largely being used in the field of geography. During 1952-57, out of 305 doctoral dissertations submitted only two were concerned with photo-geographic techniques. But it has been observed that many researchers, scholars and teachers are now using Satellite Imagery and air-photographs but they attach very little importance to it. But it has now been established that there are definite processes, methods and formulae for using Satellite Imagery and aerial photographs. As such it should not be studied haphazardly and without aims. So in this book, Satellite Imagery and air-photographs are studied for definite aims and specific purposes i.e. for studying geomorphology, landuse, urban settlements and soil geography. Before starting such studies, one should think of site selection which is one of the prime importance and to be frankly speaking that distance is one of the factors which should be given careful thought. The area selected should not be far from the Base Camp.

Before going into the details of work some elementary knowledge about the development of photogrammetry and evolution of growth concept of aerial photo-interpretation are needed. So it is clearly understood that after getting the aerial photographs, we cannot immediately to go to the field, because some amount of exercise in a planned manner should be done in the photo-laboratory to get the first hand knowledge of 'the area'. We can even explore the possibilities the collateral information which may be termed as secondary information. It can be further extended to local information during field investigation to substantiate the observation.

It may be said that during the process and development of the subject another methodology evolved since 1970s. A new technology is borne with the advent of satellites thrown into space. It is popularly termed as Remote Sensing Technique and its application is to unveil natural resources of the Earth. The term remote sensing was coined in the early 1960's by geographers in the Office of

Naval Research, U.S.A. to apply to the information derived from photographic and non-photographic instruments. These methods are also applied in this book.

The technique involves the following procedure :

(a) The general use of aerial photographs,
(b) The use which the soil sciences have in common with various other sciences, including various interpretation science and some aspects of photogrammetry.
(c) The specific use of aerial photographs in different sciences such as study of terrain, geomorphology and soils etc.

Geographers are scientists who focus on the surface phenomena their cause and effect, as they study the surface of the earth. In recent past, it is observed that geographers have often been handicapped in the persuit of their field work by considerations of time, expense, distance and weather. But now, aerial photographs are used to increase the efficiency of geographic field work. Now geographers have got two new tools, i.e.; Satellite Imagery and aerial photographs and their interpretation without physical touch constituting the basic part of remote sensing .

Briefly Remote Sensing is the technique used to obtain information about earth's surface and atmosphere without having physical contact with the object of study, usually by means of radiation from electromagnetic spectrum. Depending on it, the physical feature and chemical properties of the earth's surface emits different amounts of reflected energy derived from sun and this reflected energy produces images known as spectral signatures. The images in the form of a frame generally of 23 cm by 23 cm are available for study. The frames are studied by various types of Remote Sensing instruments which are detailed below.

When we use remotely sensed data as input, the remote sensing equipments can be broadly grouped into two categories.

(a) Visual interpretation and information transfer aids when photographic products are used as inputs.
(b) Digital image processing system when the data stored in computer compatiable tapes are used for analysis. Instruments like MDAS, NAC 100 Image Analyser, Vax 780/11 are used alongwith CCT.

Visual Interpretation and Information Transfer Aids

Based on the purpose of use, the visual interpretation and information transfer instruments can be broadly grouped into three categories :

(a) Instruments useful for viewing aerial photographs and satellite images either stereoscopically or non-stereoscopically.

    (i) Light table.
    (ii) Stereoscopes, both pocket and mirrors and scanning devices.
    (iii) Large format optical enlargers.
    (iv) Additive colour viewer.
    (v) Electronic colour combiner like image analyser with certain specific enhancement.

(b) *Measuring instruments* : Instruments useful for linear, area, height and densitometric measurement.

(c) *Transfer equipments* : Instrument that records or transfers details based on either Camera Lucida principle or employing optical projection. These instruments are used essentially for transferring the details to the required scale. Instruments such as transferoscope, stereozoom transferoscope, vertical sketchmaster uses camera lucida principle in which the eye receives two super-imposed images, one from the photograph and another from a base map, optical reflecting projector, map-o-graph, panto-graph, etc., employ principle of optical projection, where the images of photographic prints through a lens and mirrors onto a mapping surface.

Evolution

According to Robert N. Colwell (1979), the development of remote sensing can be divided into 2 time periods. Prior to 1960 aerial photography was the sole system used in Remote Sensing. But with the coming of space programme in the early 1960s, emphasis has been paid on Satellite Imagery in geographical problems. According to E.C. Barret and L.E. Curtis (1976), the geographical uses of remote sensing have co-ordinated the use of remote sensing in geography which subsequently has been designated as photo-geography. The chronological list is depicted below about the evolution of the Remote Sensing Technique for geographical uses.

(a) *Pre-1925 period :* During the First World War, in 1914 and subsequently in late 1920 some regions were being photographed systematically from the air.

(b) *1925-45 :* A period of superficial use of aerial photography, and the haphazard interpretation was mostly confined to military purposes and civilians tried to get some benefit out of it. Photographs were even attempted in hostile environment like Antarctica.

(c) *1945-55 :* Many geographers tried to be acquainted enthusiatically but its application of interpretation was in darkness .

(d) *1955-63 :* From this period the real start of photo-interpretation as a science has been developed under Dr. Ir. A.P.A. Vink (1963) and Dr. P. Burring (1960) . This period is known for wide-spread application of aerial photography.

(e) *1960 to the present day :* The gradual transformation of aerial photographs to Satellite Imagery; and scientists, researchers are now more interested about space. Geographers are particularly interested in space geography. This period is called a period of active platform and sensor experiment. On the basis of ERTS-I (1972) Dr. Peter Hagget created a model for a gateway of environmental sciences. But with coming up of new methods, in this book that model is modified in the world of modern technology now .

Growth

Remote Sensing in Pre-1960 Period

It was about 2300 years ago that Aristotle made first attempt to produce photographs. That records of images remained permanantly and the name of the Camera was "Camera Obscura." Lots of experiments have been carried out since then by such scientists.

Leonard de Vinci, Leviben Gerson, Roger Bacon, Demiel Kelper, Robert Boyte, Robert Hook, William Wallstone, George Airy and others. Their experimentation set the stage for development of photographic process, Louis Jacques, Mande Daguerre (1887-51) and Joseph Nicephore Niepce (1865-1933), they jointly announced in August 19th, 1839 their invention of the "daguerrs-o-type. The invention was announced by the French Government and the inventors were rewarded by the Government by giving them life time pension. The excess unexposed silver hallides could be removed without damaging the exposed light sensitive material. 1839 Daguerre and Niepce learned Herschel's earlier (1819) discovery that this thiosulfate chemicals could be used to fix the enclosed hallides and that fixing was key to the process.

Later optical glass was used on which the basis of modern photography was established.

During the initial stage, aerial photographs were taken from a height at about 350m from captive balloons. This was first tried in 1860 by Samuel. A. King and James W. Black who took photographs of Boston, Massachusetts from a captive balloon from a height of 350 m. At the end of the 19th century the U.S.A., U.K., France, Germany and Russia made considerable progress in this matter. During Second World War the Germans had excellent photo-reconnaisance and photo-interpretation capability. They were followed by U.K. and U.S.A. In 1927, use of aerial photographs was advocated in the study of landforms and soil Surveys in U.S.A. , followed by efforts at photo-interpretation within a few years by Bushnell, 1932. Simultaneously, it was taken up in Australia by Prescott and Taylor (1930) and in the USSR by Levenhampt (1930). Use of aerial photographs by the defence personnel was made by various nations of the world, particularly during Second World War. In 1945, the U.S.A. Government has brought out an excellent monograph entitled Photographic Intelligence Centre. Now before entering into the space science, some attention is needed to the development of aerial photographs. The scientific use of aerial photographs contains various aspects which need careful definition. Lack of definition has seriously hampered the thrust areas.

The following questions naturally arises in this context

(i) When to use aerial photographs ?
(ii) How to use them ?
(iii) To what purpose ?

A problem of particular importance in this respect is the combination of the information gained from aerial photographs with the indispensible field observation. This is the central question to which many different answers are given by different persons. In particular in this field misunderstanding occurs. This is partly due to vague definition and partly due to the different circumstances and the purposes for which the aerial photographs are used. A very critical approach to all literature is necessary in this respect.

Landform, particularly the micro relief could be well studied through aerial photographs. The soil is not, however, seen on the photographs to any extent, sufficient for description, but if the land is open, some amount of inference could be drawn by the retention capacity of water in depression and natural stream valleys. Tentative soil map thus prepared, could be verified. So field control of soil investigation, therefore, is of much fundamental importance than superficially comparable 'ground control' , of photogrammetry and some of the other uses of aerial photographs.

It has been ascertained that the aerial photographs are used in two ways, viz. (1) general use, and (2) specific use. For general use, these photographs are of utmost necessary because scanning of aerial photographs is necessary which is the first step of interpretation. A geographer getting aerial photographs may draw general conclusion as the location of a place in the tropics, humid tropics, arid or semiarid region where many indications are reflected on the aerial photographs. All these elements on the surface of the earth starting from water bodies, vegetation, landuse etc. are closely marked on the photographs and finally a conclusion could be drawn on the photographs supplied.

So general use of aerial photographs are of the nature of interpreting topographical sheets because, it does not include specific purpose. It is not true, however, to say that in this case no photo-interpretation is done, as the use of aerial photographs either by photogrammetric method or by interpreter is based on interpretation of his/her'visual observation in the photogrammetric image. This is clearly demonstrated by the way in which 'photo-interpretation' has been used by a recent Russian author (Goldman, 1960). In this case, indentification and recognition of objects in topographic maps is also considered. Therefore, all users of aerial photographs offer certain common interpretation. This should be called the general use of aerial photographs. So the use of photo in photogrammetry is very close to this general use in its fundamental aspect but the specific mathematical approach certainly has large domain of specific use in photogrammetry.

But now it should be described what is interpretation science, which is completely different from general use. The specific use of aerial photographs in this context is often called as the "Interpretation Science" and it also includes various aspects.

The Commission VII of The Informational Society for photogrammetry (Archives 1962) gives stress on —

(1) Photography, equipment and techniques,
(2) Geology,
(3) Soils,
(4) Vegetation,
(5) Regional Geography and Planning,
(6) Ice,
(7) Archaeology,
(8) Oceanography and Coastal Research, and
(9) Engineering.

In this dissertation Item No. 1, 2, 3, 4, 5 have been taken into account. The photo-interpretation follow the following sequences.

(1) Recognition and identification
(2) Analysis
(3) Deduction
(4) Classification and idealism

The above mentioned procedures are still valid for visual interpretation of aerial photography and satellite imagery. But the method is completely different as in case of aerial photographs a fine model could be created but such model is not possible in satellite imagery. Grey level values could be identified by visual method and separated the pixel values (picture element) .

Recognition and Identification

This is the first phase of photo-interpretation which consists in obtaining from the photo a clear and complete picture of ground condition beginning with the identification of individual features-hills and valleys, streams, roads, buildings etc. ending with the integration of scattered observation into a comprehensive view of the terrain as a whole with all parts in proper perspective. Collateral information of land such as maps, reports, and local observation would be of great help to evaluate the study and initiate new formule regarding the transfer of technology. Burring (1960) makes a distiction between the identification of cultural objects like houses, roads, canals, bridges, reservoirs, afforestation plot etc. which is rather simple, while identification of land units like terraces, escarpments, U-shaped valley, fault, thrust etc. which is more difficult. This type of study is only possible for trained personnel in aerial photographs.

There are two methods of air photo-interpretation, viz.

(1) Fishing Expedition.
(2) Logical Search

Before drawing planimetric data, contours or map units or contouring objects, the interpreter should delineate 'effective areas' on the photographs. The effective area of a photograph includes all images closer to the centre than to the centre of any other. It is the area in which objects can be viewed with least relief displacement. The effective area is the central part of aerial photographs and it is around the principal point of geometric centre of photographs where the linear displacement is almost nil. In vertical aerial photographs the areas are well presented. In the aerial photographs of the mountains or uneven areas, relief displacement is inevitable. This is usually corrected by the relief displacement formule.

Relief Displacement

It is the distance between the position of a point on the photograph and its actual position due to relief.

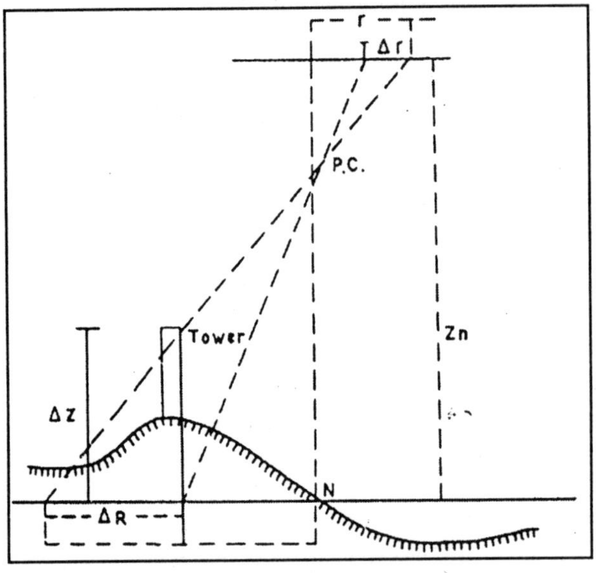

FIG. 2.1 : *Relief Displacement*

Source : International Training Centre for Aerial Survey (ITC) Chapter B IX 1960

In the diagram —

$$\frac{\Delta R}{R} \qquad \frac{\Delta Z}{ZN} = \frac{\Delta r}{r}$$

Thus, $\Delta r = r \dfrac{Z}{ZN}$, in which

Δr = Relief displacement
ΔZ = Height difference over/under reference plane
ZN = Flying height above reference plane (chosen through Nadir point)

It is therefore evident that relief displacement increases from the Nadir point.

Single-lens vertical aerial photographs are used as base map in soil survey, land classification and soil conservation investigations. The aerial photographs or the photo-mosaic replace cadestral topographical or geographical maps. Field investigations can be carried out in an efficient way and both speed and the accuracy of the work can be assured. In Fishing Expedition the interpreter gets a detailed idea of the terrain condition and develop the catena concept. It can be said that recognition and identification of lands and terrains are beginning of the interpretation science for the preparation of map. The final product is the map with literature for understanding the maps. At the end of the recognition and identification one should prepare a photo-key for the purpose. The photo-key is the basic requirement of an interpreter what he could extract from a stereopair, It is therefore, safe to identify in a limited way comparable to photo reading of Laueder (1959) . The photo-key is very important in soil survey as well as efficient planning in the field (Vink, 1963).

Air Photo Analysis

Laueder (1959) describes this is "the process of separating anything into its constituent parts or examination of anything to distinguish its component parts separately or in their relation to whole to be applied to the various features of aerial photographs". It includes all the aspects of the photo-reading but also adds that this is an evaluation of the numeric inter-relationship of photo features. Systematic analysis is the main feature of the analysis. It actually means reservation of elements separately as per desire of the worker and then to draw bounderies. In photo-analysis the establishment of individuality is carried out in the stereo-image and the determination of identity is fully left to field work. The field work is also needed to correct possible mistakes in observation. Photo-analysis only gives lines which indicate probable difference of soils without trying to compare the various forms and surface at any distance.

Physiographic Classification of Air-Photo Interpretation

Physiographic classification is the main criteria of landform studies. Landform studies could be effectively utilized on the photo-image which could be easily done by a good interpreter. In the systematic analysis, one or more elements are presented as terrain units. Subsequently, these units are defined and classified in terms of analogue. Different or varying physical or cultural characteristics of the earth's surface are recognised on aerial photographs. This type of air-photo interpretation has been developed in order to collect data of a composite nature for classes.

Investigation of terrain condition is the first step of soil mapping. Instead of analysing a number of single elements, these terrain units are directly identified, analysed and classified according to the specific characteristics as shown in the photo stereo-image and accordingly to deduct about physiographies which have been generally found to correlate with the different soil units.

The second approach is more difficult as field study is essential for making collateral identity. The physiographic classification of photo-geology is an, important step in aerial photo-interpretation. Careful observation, analysis and clarification based on these elements can ensure accurate definition and measurement. In this way the procedure of air-photo interpretation is most helpful for mapping terrain features. It is considered that about 10 per cent of the mapping area should be carried out to the field. This 10 per cent should be marked in on air-photographs in the overlays in such a way that it covers all elements by which the expert interpreter interpolate and extrapolate in the photographs and delineate the boundaries in the laboratory. So the method of conventional field work could be reduced by 80 per cent. This is the advantage of air-photograph and its interpretation. Systematic analysis of all the important elements is the basic criteria of photo-interpretation.

The basic elements are :

(a) Grade or density (degree of slope, gradient etc.)
(b) Shape
(c) Size
(d) Regularity
(e) Site or geographical position

Special Indications are :

(a) Distinction (Contrast)
(b) Shadow
(c) Colour tone
(d) Colour texture

The elements are categorised and classified as follows :

I. *Elements :* General morphology, land type relief form, slope, alluvium plains, pediments, drainage pattern, watershed pattern, river-cum-creek pattern, lagoons, bars, etc.
II. *Elements :* Special aspects which are essential to pinpoint the objects of study like gullies; colour contrast in grey level values indicates water and streamflows.
III. *Elements :* Natural vegetation, crown densities and its texture and identification of species of plants, landuse.
IV. *Elements :* Man made — dams, reservoirs, canals, roads, dykes, agricultural land, urban areas, archeological objects.

It may be said that thorough dissection of these elements are necessary to obtain the perfect knowledge of the land. There is no strict line of direction between analysis of individual element and physiography. So expert photo-geographer tries to correlate the individual element with physiographic elements. This could be matched. Some examples are cited here to make things clear.

1. *Combination :* gully — escarpment, alluvial flat typically observed at the junction of Terai and

Bhabar region of Dehra Dun at Raja-Ji Game Sanctuary.

2. *Terrace escarpments :* (bluffs) — *cum*-terrace scars which are clearly observed along Rishikesh-Hardwar Road near Rishikesh; low terraces of the Ganga are cut for road 'linkage.

Elements

1. Land type is the largest unit of interpretation. It is nothing but to segment out local physiographic unit of a larger one and in this, it is to be cleared that one landsat imagery covers 34,225 sq.km. Where as an air-photograph of 1,25,000 scale covers 134 sq. km. Only one satellite imagery contains near about 400 air-photographs. So a satellite imagery gives more synoptic coverage and it can be scanned in a very big wider aspect for 400 air-photographs of the scale 1:25,000. So from air-photographs relief map can be prepared of smaller area at a time, whereas, general and broad relief forms could be ascertained in satellite imagery. The main purpose is to choose the right type imagery. But to get first hand information Satellite Imagery are useful and aerial photography could be effectively utilized by researchers.

(a) *Grade or distinction :* This is very important for variations to be mapped. Descriptive names and classification should be made in standard guide line to be used by physiographers and geomorphologist or by analytical elements of slope and landforms.

(b) *Shape :* Soft or rounded, sharp land-types

(c) *Size :* Macro, Meso, Micro land types.

(d) *Regularity*

(e) *Site :* The site chosen is important and depends upon the judgement of the researcher. Both land type and relief-form together are sometimes called land forms. In Dehra Dun much emphasis has been paid on the land form because, it is located at the foot hills of the Himalayas forming a part of the Dun Valley. At first detail studies were carried out; on the basis of aerial / photographs in the scale of 1:25,000. Subsequently, satellite imagery of 1:1 million scale had been consulted to cover bigger area and to elucidate the broader aspects. In this process, the photo-geological and landform map of the same area have been prepared.

2. *Relief form :* (a) The stereo image study of aerial photographs are very helpful to classify landform (relief) types, as well as deciphering the grade (c) site already mentioned earlier.

In photogrammetry, relief forms are mapped by the parallax bar and linear displacement formulae. Any displacement can be adjusted by factors as the aerial photograph is a central perspective projection and relief form are made by form lines .

3. *Slope :* (a) This is an important criterion as there is standard stereoscopic formulae which is detailed below :

The term parallax is applied to the apparent change in the position of an object caused by change of position in the observer. Points at higher elevation exhibit a greater parallax than-those at lower elevation.

O_1 and O_2 is the air base B at a vertical distance of ZA and ZQ above the terrain points A and Q respectively; a_1 and a_2 are the corresponding images of object point A on the photograph. Focal length of aerial camera lens is f.

From O_2 draw a line O_2 a, parallel to O_1q_1. Then by definition parallax of A is

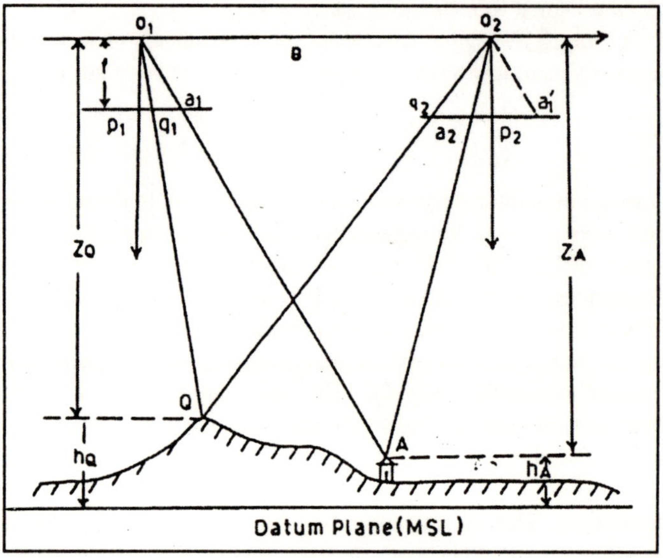

FIG. 2.2 : *Geometry of Parallax*

Source : ITC Chapter B IX 1960

$PA = a_2 a_1$

From similar triangles $O_2\, a_2 a'_1$ and AO_2O_1

$$\frac{ZA}{f} = \frac{B}{a_2 a_1} = \frac{B}{P_p}$$

$$\therefore ZA = \frac{B.f}{P_A} \qquad \text{.....(1)}$$

Similarly for a point Q, $Z_a = \dfrac{B.f}{P_Q}$ .....(2)

From equation (1) & (2) $Z_A - Z_a = \dfrac{B.f}{P_A} - \dfrac{B.f}{P_Q}$

$$\therefore ZA - ZQ = Bf. \left(\frac{1}{PA} - \frac{1}{PQ}\right)$$

$$ZA - ZQ = P.f\left(\frac{1}{PA} - \frac{1}{PQ}\right) = Bf.\ \frac{P_Q - P_A}{P_A\, P_Q}$$

$$= \frac{B.f.}{PA} = \frac{P_Q - P_A}{P_Q}$$

$$= ZA \,.\, \frac{P_Q - P_A}{P_Q} \quad \text{.... by Substituting for } B.f/PA \text{ from equation (1)}$$

$$= ZA \,.\, \frac{P}{PA + P} \quad \text{.... (3)}$$

where $\Delta P = PQ - PA$.

Let h_A and H_Q be the mean sea level heights of object points A and Q respectively. Then $ZA + hA =$ Flying height of the aircraft above datum plane (MSL)

$$= ZQ + hQ.$$

$$ZA - ZQ = hQ - hA$$

= Difference of heights between terrain points Q and A

$$= \Delta h.$$

Equation (3) now can be written as —

$$\Delta h = \frac{\Delta ZA \,.\, \Delta P}{PA + \Delta P}$$

All the air-photographs are perspective pictures and various types of planimetric inaccuracy are contained within them. Overlapping photographs are capable of providing stereoscopic image. When a photograph is vertical the isocentre and Nadir point are coincident with the principal point. The stereoscopic parallax can be measured with the help of parallax bar (an instrument) and from the parallax of two different places parallax difference can be obtained. From the parallax difference (ΔP) height determination is possible. In the formulae stated above, the difference between these two parallax measurement arises from the height difference bstween A and B, therefore difference in parallax can be used to calculate height differences from airphotos. Besides these there are other methods such as Slotted Template Method and Radial line plotters. The last two instruments are not generally available. So parallax bar is generally used for height determination. It is not confined in the principal points of the stereo-pair.

So by the relief displacement and parallax formulae one can determine (a) slope gradient (b) slope length (c) slope regularity (d) site etc.

4. *Drainage Pattern :* This item can be easily interpreted by the geographers particularly by photo-geographers. Things to be observed are :

(a) Occurrence and density
(b) type (c) size (d) regularity or anomalies any in unnatural change in camera.

5. *Watershed Pattern :* (a) density (b) type (c) size (d) regularity (e) site.

6. Rivers, streams, creeks (patterns and individual), density, type, shape, size, regularity, site.

Elements related to special aspects of the tarrain

7. Stratigraphy (a) gradient strike (b) horizontal, concave or convex stratigraphy (c) size, thickness of layers or particular beds. (d) regularity (e) site. .
8. *Gully form :* density, type, shape, size, regularity and site.
9. *Gully pattern*
10. *Colour of the Earth's Surface :* (a) tone or colour intensity. This has been changed recently which is known as spectral signatures. Still texture is very important like stoniness, rockiness, soil (rust, sheet, erosion, size, regularity, size, etc.)
11. *Water and Drainage condition :* dry, marshy, flooded areas, area covered by floods, etc.

Elements Ralated to Vegetation Cover : Natural vegetation, or afforestation, social forestry etc., type, size species, crown density etc. or density of forest like dense forest, thick forest.

Most of the virgin forest of the country is almost lost. The primary and secondary vegetation could be ascertained in aerial photographs as for example in Surinam secondary forest is more dense and they are the main source of timber than primary forest. The main cause is application of chemicals, which accelarate growth of forest.

12. *Specific trees :* Tropical rainforest, Temperate forest and Arctic vegetation etc.
13. *Landuse :* For landuse mapping, aerial photographs is most suitable to get up-to-date information about the use of land. The landuse map is prepared by different mapping organisations and they make their own classification and it is not possible to interpret all these as mentioned in the classification. As for example in air-photographs we can delineate the arable land boundaries but cannot identify the crop combination as the air-photographs are taken from the height of 9150 m from the sea level. Of course, low flight air photographs solve some of the problems. So the crop combination is generally imposed on the arable land crop combination can be made from the statistics available from local Panchayat office. The following can be identified in aerial photographs and satellite imagery (1) Forest (2) Arable land (3) Wasteland (4) Urban sites (5) Bad land such as gullies (6) Water bodies which includes tanks, ponds, rivers and streams (7) Brick fields (8) Industrial sites and any other objects which could be identified by collateral information.

Elements Related to Human Aspects

14. *Ditches and canals :* Density type size — width, depth, length.
15. *Dykes :* Big canals coming out from the reservoir, as for example two big canals which comes out from Hirakud reservoir of India known as Bargarh canal and Sason canal. The dykes of the reservoir, the dykes of the Hirakud reservoir are 25.8 km. long, supposed to be largest dyke was constructed in India, because the Hirakud reservoir forming the biggest artificial lake in Asia with a reservoir spread of 737 sq.km. irrigating 264,000 hec of land in Sambalpur and Bolangir Districts of Orissa of India.
16. *Parcelling :* Percelling of land is very prominent in the arable land. This is typically known as fragmentation of holding. This is noticeable in the satellite imagery but in aerial photographs it is quite prominent.
17. *Variation identity,* of types size, regularity and site of roads in particular of the older roads and of rural roads and the road pattern, often indicates quite different terrain conditions. In photo analysis only the establishment of the individuality is carried out in the stereo image and the determination of identity is fully left the field check.

Deductions

Lueder (1959) uses deductions as normal part of "Photo-interpretation". Buring (1960) is not interested to use this word in soil science. He insisted that photo-interpretation is guess work, whereas soil survey is not. He only allows deduction in two of his eight procedures of soil survey, i.e. interpolation or extrapolation.

In deduction the most important thing is the 'reference level of interpretation' . The reference level of interpretation of one may be completely different from another researcher. So it is standarised by the photo key which makes a frame and all that requirements are included in the frame.

Aerial photo deduction deals with the combination of careful observation on aerial photographs and knowledge from other sources in order to obtain information that cannot directly be observed in the photo image itself. This, howover, is still not always enough. In soil . survey at least the field and laboratory checks are needed (Buring 1960).

Classification and Finalisation

This is the last stage of the photo-interpretation and it gives every thing we desire such map, legend, reference and to go for press to obtain the desired copies for sell in the market. But before taking any decision a field survey is necessary to bring final checkings. This last step is called photo-interpretation by Lueder (1959). It is preferable however to use the term photo-interpretation in the wider sense which is common (Archives 1962) and which also include the process of photo-reading and photo analysis. In fact each of them also includes some classification of individual objects in the former. Finalization is even more general because any map includes some finalization of the natural pattern as depicted in the photo image (Vander Weele, 1963).

At this stage, the researcher tries to delineate the lines of probability.

(1) Continuous line to indicate high reliability.
(2) Dotted line to indicate relatively low reliability.

Requirements of Photo-geographers are as follows :

(i) Few copies of the Photo Index.
(ii) A good number are required contact stereopairs of photographs.
(iii) Photo-mosaics preferably in semimatt papers are required. Scale 1:25,000 or any other desired scale.
(iv) Sets semimatt or matt contact print in 1:25,000 scale or any other scale are required.
(v) One Mirror stereoscope.
(vi) One pocket stereoscope and comparable instrument.
(vii) Scanning stereoscope, old delft.
(viii) Sketch Master and radial line photters.
(ix) Stereo-pantometer to draw the contour lines.

Of course the above mentioned instruments are available in normal photo-interpretation laboratory. But the procurement of aerial photographs is not so easy. If the choice of the area is in the declassified zone it is more easy. But it requires some procedure to procure the air-photograph from Airforce through Survey of India. Any researcher, who wants to do work on photo-geography, will get the necessary instruction from the agency. After obtaining the necessary facilities of getting the instruments, we may go into the process of photo-geography to prepare soil from the air- photographs. The systematic aerial photo interpretation follows the sequence out lined below :

Group A : Procedures of most conventional field work.
- (i) the field map procedure
- (ii) the brief orientation procedure.

Group B : Procedure of photo-interpretation of largely conventional field work.
- (iii) the photo preparation procedure.
- (iv) the adjusted interpretation procedure.

Group C : Procedure with partly adjusted and fully controlled photo-interpretation.
- (v) the pre-interpretation reconnaissance procedure.
- (vi) the interpretation with full check procedure.

Group D : Procedures of partly controlled photo-interpretation.
- (vii) the interpolation procedure
- (viii) the extrapolation procedure.

Aerial Photo-interpretation in various kinds of Soil Surveys

- I. Reconnaissance Soil Survey
- II. Detailed Soil Survey
- III. Semi-Detailed Soil Survey

The five methods of soil survey are numbered ABCD and E they represent (Fig. 2.3).

A. A reconnaissance survey carried out according to the standard of U.S. Bureau of reclamation. One field observation is made on every 100 ha. in a grid system.

B. A detailed survey carried out according to the standard of U.S. Bureau of reclamation. One field observation is made on every 25 ha. in a grid system.

C. Semi-detailed soil survey carried out according to conventional standard for complete field survey with one observation per 4 ha. using aerial photographs as field surveys with one observation per 4 ha. using aerial photographs as field surveys with one observation per 4 ha. using aerial photographs as field and working map.

D. A semi-detailed soil survey, based on combination of field observation and aerial photo interpretation according to procedure V. An average of observation and aerial photo interpretation per 10 ha. is made.

E. A semi-detailed soil survey based on the method of interpolation procedure VIII.

Remote Sensing after 1960

Since 1960, there occurs considerable changes in Remote Sensing. The earlier conventional methods of Remote Sensing were based on aerial and terrestrial photographs. Recently Remote Sensors are placed within the satellite particularly on lunar orbiting satellites so that the manned lunar space craft could be safely landed. Systematic earth orbital observation by United States Satellite began in

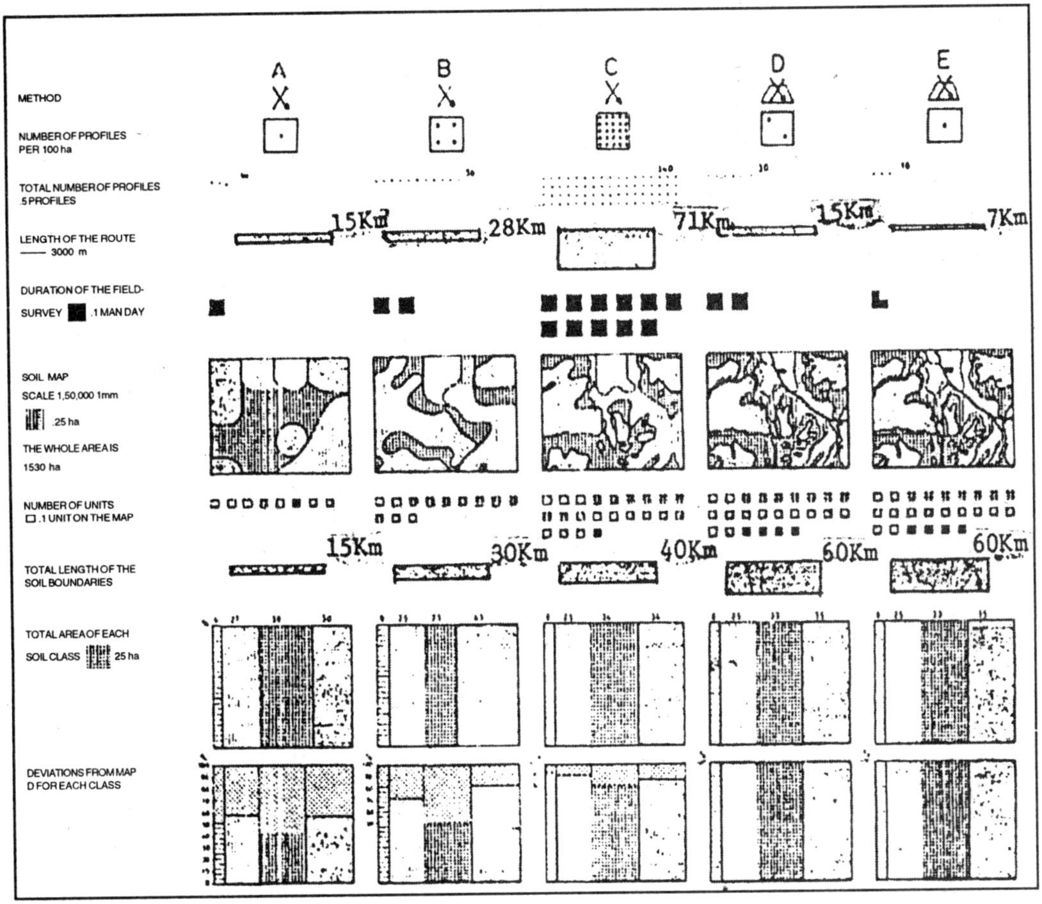

FIG. 2.3 : *Comparison of the Results of Five Methods of Soil Survey*

Source : Burring, Five different methods of soil survey, 1960.

1960 with the launch of Television Infrared Observation Satellite (TIROS-I) . TIROS-I was the first meteorological satellite and carried a low resolution imaging system. Since the launch of TIROS-I. NASA has landed more than 40 meteorological environmental satellites with steadily improving sensor data collection capabilities. TIROS satellites were originally supported to Environmental Science Service Administration (ESSA) which was absorbed into the National Oceanic and Atmospheric Administration (NOAA) in October (1977).

The following sensor systems are placed in the satellite —

(1) The advanced very high resolution radiometer (AVRR).
(2) TIROS — This is a package of three sensors, a High Resolution Infra-red sounder, a stratosphere sounding unit and microwave sounding unit.
(3) The data collection and platform location system (DCS) .
(4) *The Space Environ Monitor*. The device consists of four detector systems and a data processing unit. The Space Environment Monitor system for measuring proton and electron flux near the earth.

The first space photograph of the earth was transmitted by Explorer in August 1959. In 1960 the first orbital photograph was also available. Several hundred 70 mm colour photographs were taken by an automatic camera in the unmanned MA-4 [Mercury Atlas) Space craft to take photography of Earth's surface. J.A. Mc. Divitt and E.H. White II conducted the first formal geologic photography experiment on the G.T. 4 (Gemini Titan) Mission June 3-7, 1965 (Lowman 1967). The four day mission produced a spectacular service of 39 overlapping nearly vertical photographs of the South-Western United States and Northern Mexico. The Mission demonstrated that useful data could be acquired for earth resource analysis. Besides the USA sends a number of satellite and obtains valuable photographs. The American astronauts used a series of four electrically driven vertically mounted 70 mm camera, to photograph the terrain with four different film/filter combinations — panchromatic film with red and green filters, black and white *infra-red* film and colour *infra-red* film . The spectral bands employed were close to those planned for the NASA Earth Resources Technology Satellite (ERTS-1). Over a period of four days, the crew obtained 350 pictures of good quality. The whole process of sending satellite in the space was organised by the scientists of the NASA and one in India was also involved in it. This is the age of interpretation of landsat. Now we must know how LANDSAT came into existence and subsequently the whole market of the world is captured by NASA and new mothodology is evolvod for image analysis system and many modern instruments came into existence and Japan and Eastern Germany took the lead to interpret those satellite imageries either by digitization or analogue interpretation by instrument such as, Image Analysis 100 invented by TOKYO University. MINATOKU under the leadership of Dr. S. Murai who is the Secretary of Asian Remote Sensing Conference. I met him in the 5th Asian Remote Sensing Conference when I presented a paper in Colombo in 1984 and discussed with him the future development of the Remote Sensing Technique in India. Japan has developed good instruments for image analysis which shows not only the area per frame but histograms are depicted indicating the proportionately different categories of 15 colour frequencies. In the USA the first satellite designed specifically to collect data of the earth's surface and resources was Earth Resource Technology Satellite (ERTS-1). The first three of this series of modified Nimbus Satellites were launched into near — polar circular orbits at an altitude which varied between 897 and 918 km. The ERTS was later named as LANDSAT-1. Uptil now four LANDSATS have been launched in the following order :

(1) LANDSAT-1, July 23, 1972, four channel Multispectral scanner (MSS), a three camera return beam vidicon (RBV), a data collected system and two video tape recorders. The MSS operated in the following spectral intervals : Band 4 (0.5 - 0.6μ) Band 5 (0.6 - 0.7μ).

Beside these 3 independant cameras of RBV covered three spectral bands, that is blue-green (0.47-0.575μ), yellow-red (0.58-0.68 μ) and near infrared (0.69-0.83 μ). Both of these system viewed a ground scene of approximately 185 km: by 185 in area, with a ground resolution of 80 meters. Later RBV was abandoned due to technical difficulties.

(2) LANDSAT-2, January 22, 1975. This satellite carried a load similar to that of LANDSAT-1.

LANDSAT-3, March 5, 1978. This satellite added a 5th band in the thermal infra-red (10.4-12.6μ) to the MSS. The RBV cover system was also significantly changed. It carried for experimental basis covering the spectral band from 0.53 to 0.75 μ. It was designed simultaneously to view 84 sq. km. of ground segment, giving a linear swath of about 185 km. Its resolution of 40 meters was better by a factor of 2 than that of previous MSS & RBV system.

Now there is again a change in codification, LANDSAT 1,2,3,4 now called LANDSAT A,B,C,D.

LANDSAT-D adds another band which is called Thematic Mapper and it was bunched. Thematic mapper has unique features as it contains 96 detectors by 7 spectral bands having resolution of 30 meters and 7th band gives resolution of 120 meters. MSS could scan in one direction, whereas, the TM could scan in two directions. The TM detector arrays are located in the instrument's primary focal plane and thus incoming radiation to be reflected directly onto the detector, while in the MSS the incoming light is transmitted through fibre optics before being reflected on to the detection arrays, with some loss and degradation of the signal. Major and important inprovements in data acquisition have achieved though the use of series of communication satellites (DOMSAT). When the Landsat vehicle is within range of a ground receiving station its data will be transmitted directly. MSS data will be transmitted at 15 megabites/second on 'S' band and T.M. data will be transmitted 85 megabites/Second on ' X' band.

Now the process is completely changed in different parts of the Earth. The NASA has established the ground station, when the satellite passes over the ground station, they relay the sorted material to the ground control station and the digitalized data are collected in Ampex magnetic computer tapes. As in India the ground control system is located at a distance of about 110 km from Hyderabad city. They collect the data every day in 6 Ampex magnetic tapes. By the quick look process, they scan the whole of the portion it covers. Of course, regarding time and data, satellite passes chart that is being sent to the each ground station of the world fortnightly by NASA. From this, we get the timing of the satellite movement around the world. A fresh antena is constructed in Sahdnagar (Hyderabad) to catch the SPOT data.

In the history of Landsat Imagery Thematic Mapper (TM), an enlarged T.M. is as good as a map if properly annotated. In the meantime, many countries like USSR, Germany tried to launch their satellite, but it is very difficult to construct launching pad, only two countries have launching pad station, one in U.S.A. in Houston Goddard Space Flight Centre (GSFC) and other is Russian Cosmodrome in Baikanur in Siberia. Besides experimentation, many types of launch vehicle were introduced in the space, Such as Sky Lab, HCMM, Sea SAT, Space Shuttle etc. It is not possible to give detailed description of these projects, because they do not possess in continuity, only for experimental purpose. The space Shuttle of DFVLR of FRG is very promising device. The method relies on stereo pair of photographs during the flight mission and can be interpreted by the stereoscope, but shuttle flight are very limited particularly in its area. Russian Shuttle flights which include one of the two Indians on the space during its flight of the 260/70 km. above the surface of the covering parts of Orissa and of Kutch in Gujarat. The picturesque element is easy to interpret.

SPOT in France is also trying to come forward and the European Space Agency (ESA) has developed an expendable launch vehicle called ARIANE in France which is capable of delivering a play-load of 1700 kg. into a geo-stationary orbit. It is launched in April 1984 from the French System Probatoired Observation *de la Terre* (SPOT). It is an earth observation satellite which will have finer ground resolution than any other LANDSAT series to date (1, 2, 3, D). The Satellite will be placed in Sun Synchronous Orbit, i.e., polar to polar flight. It is required to mention that there are two types of Satellite. One is Sun Synochronous and other is Geo-Stationary Satellite. Polar to polar flight is scattered and sun synchronous. Geo-Stationary relates to latitudinal placement of satellite along the direction of Earth's rotation. The speed is the same so it is placed in a particular place above the surface of the Earth. This type of satellite is mainly for meteorological purpose and it is called Metsat.

SPOT Satellite consists of two identical High Resolution Visible (HRV) imaging instruments, two negative tape recorders and telemetry transmitter. The SPOT produces two types of images which are stereopair. The coloured one has 10 meters resolution and the black and white panchromatic mode with a ground resolution of 8 meters. The SPOT system technology shows two — distinct disadvantages ; (i) Each detector in a linear array must be individually calibrated in order to produce a uniform response over an image scene; (ii) The detectors presently available are not suitable for use, with wave layers longer than about 1.05 micrometer. The flying heights varies between 815 and 829 km.

Technology Transfer

From the evolutionary account of Remote Sensing the question naturally arises how to use those inventions and discoveries in our country and in other developing countries. Persons who are interested from the developing countries after learning about these techniques could not properly utilize their expertise due to non-availability of infrastructure in these countries. So most of the foreign trained personnel go back to the developed countries. Transfer of technology to the developing countries is the most vital thing. The agency representation often visit our country. Their only purpose is to sell the instruments, and to delay the services of the instruments. So whatever comes out we must take our own initiative as in space programme: India has advanced a lot in her space programme. The experts of SAC are quite compitent to develop and design the satellite alongwith special application of creating infrastructure. But to develop a launching station of a Satellite of giant size is not possible in our country at present. To launch a big satellite of Landsat type, requires million tonnes of nuclear energy. The nuclear energy facilities are not available in our country. Our atomic energy is only used in peaceful purposes, i.e. for the production of energy. The trend toward economic development for many Third World nations is greatly influenced by a number of fundamentally interconnected factors as geographic location, political and economic system, social structure, cultural heritage, natural and human resources. Technology transfer may be defined as the process by which scientific knowledge and skill to apply that knowledge are transferred to others. Technology transfer was once thought to be straight forward process, in which technological advances would be recognised and quickly incorporated alongwith similar needs of others (Wagner & Lawe 1978) . As it is known that remote sensing has such a wide range of potential application and highly sophisticated components, the technology should be transferred to other societies and nations with utmost care. Sometimes it has been observed that transfer of technology is not always beneficial but it may also be destructive (Wagner & Lawe 1978) . Remote sensing is a powerful tool by means of which developing countries can obtain the information needed to manage their resources. In many instances, developing countries may have the advantage in accepting remote sensing approaches. At present they are not at par with the developed countries with established system of data acquisition, analysis and management. For this reason Agency for International Development (AID) in 1970 began exploring the possibilities to cater to the needs at per National level.

Illustration of the Map

The map shows Land Receiving Station now operating or under construction (black circle) and those which are planned for near future (white circle), (Paul 8 Mascarenhas 1981). The circles represent the land reception range from receiving station (Approximately 2,700 km).

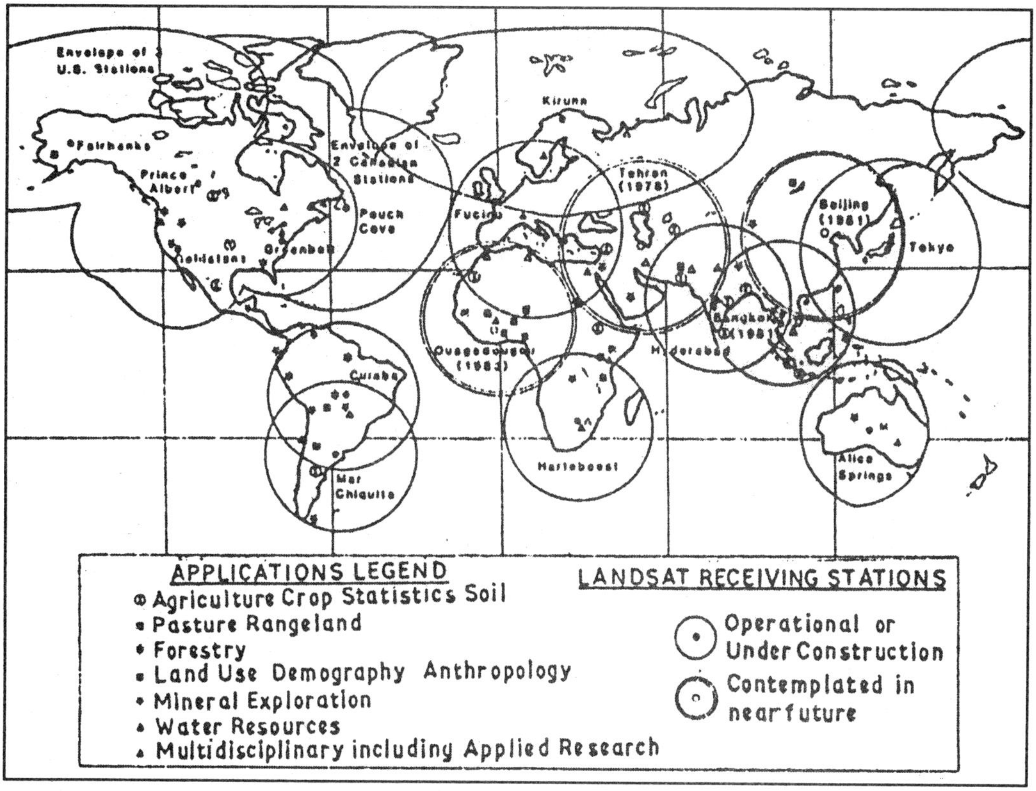

FIG. 2.4 : *LANDSAT Operating Stations*
Source : Manual of Remote Sensing Am. Soc. Photo, 1983
Note : Ilustration of Landsat receiving stations that are operational, under construction, or planned. Circle size indicates the range within which the station can receive Landsat data.

Remote Sensing is found to be useful in developing countries. Example of application include —

(1) *Landuse planning :* Identifying environmental factors which have potential effect on land cover and evaluation of land cover change through the use of Landsat data (Santos 1980).

(2) Forestry (a) Identification of degraded pasture-land after deforestation and the identification of pine and eucalyptus plantation in Brazil particularly in Gramado area (Fitho, 1980) (b) Identification of mangrove vegetation and forest resources and (wet lands for pisciculture) in Bangladesh and West Bengal in India.

(3) Geological mapping through the use of aerial photographs (Junchaeo, 1980).

(4) Identification of major geologic rock bodies using Landsat data.

(5) Mapping of soils using Landsat and ground survey (Siefferman, 1980).

(6) Use of Remote Sensing Technique for highway engineering surveys and highway planning (Beaumont 1978 and 1980).

(7) Use of Satellite mosaics to find out giant rift valley in Botswana region (Banerjee, F.AO,1985 in FRG) and other micromorphological features. Almost 120 countries have purchased landsat data from the EROS data centre and NRSA (Hydrabad). Of these roughly the two-

thirds are of the developing countries.

At the end of transfer of Technology, the following observations, are made : (i) how to introduce the remote sensing study in our country other than aerial photo-geography? These are in complete control of the Air Force. (i) They can independently carry out jobs efficiently except the procurement of camera to be purchased from foreign countries. Latest camera for high altitutde is known Metric Camera designed by West Germany in The German Aerospace Research Establishment (DFVLR) in Oberffenhoven. This Camera provides stereopair of photographs, minimising the cost of research work.

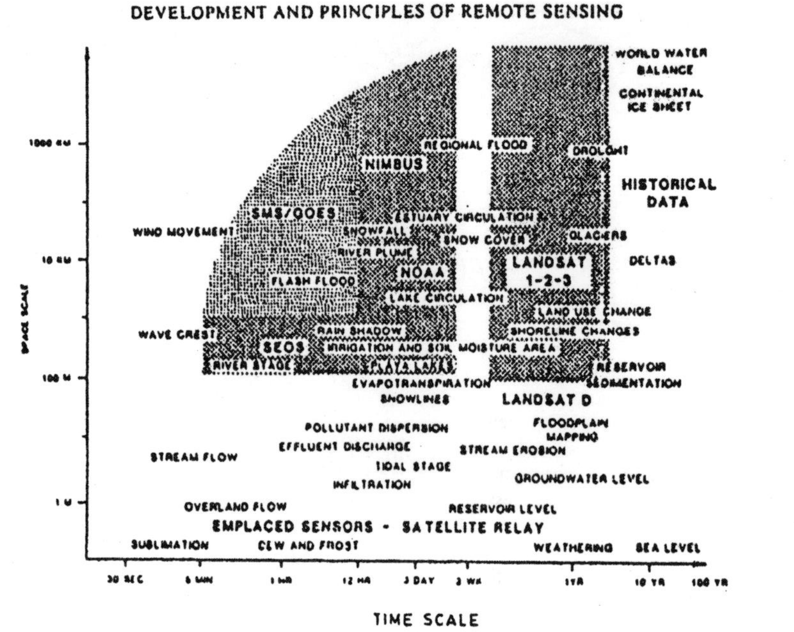

Time and space scales for representative hydrologic parameters with relevant observational capabilities of satellite systems represented by shaded areas.

FIG. 2.5 : *Development and Principles of Remote Sensing*
Source : Same as Fig. 2.4)

Principle of Transfer of Technology

The general principal of technology transfer may be as follows :

(i) A Remote Sensing Centre should have trained personnel with concentration of scientific skills.

(ii) The interaction should be made through the data analysis among personnel with diverse disciplinary training.

(iii) There should be common understanding between agency producing it and the users for the flow of data and requirement. As for example, while handing RBV data, the RBV may be inoperative due to technical flaws and the whole project is differently oriented.

(iv) The remote sensing studies must examine ground truth, as supervised and unsupervised classification.

Table 2.1 : Estimates of the Resolving Power of LANDSAT MSS

	Resolution Measure	Source	Resolution Meters
1.	IFOV - Geometric	NASA 1972	79
2.	IFOV - Gemoetric	Slater 1972	76.2
3.	IFOV - Gemoetric	Colvo Coresses 1979	73.4
4.	IFOV - Gemoetric (Min. Altitude)	Gordon 1980	76
5.	IFOV - Gemoetric (Min. Altitude)	Gordon 1980	81
6.	Pixel Size	General Electric (undated)	79 x 56
7.	Pixel size - resampled Landsat 3 CCT's	Holkenbrink, 1978	57 x 57
8.	EIFOV - Point Spread	Landgrebe *et al.*, 1977	90
9.	EIFOV - half cycle	Welch 1977	66
10.	EIFOV - half cycle	Welch 1977	135
11.	ERE	Volvo Coresses 1979b	87
12.	Modified ERE estimate for channel 4	NORWOOD 1974	125
13.	Mini classifiable area	Shay 1975 General Electronic	500 x 350 320 x 220

Source : Remote Sensing Manual Vol. I. 1983

IFOV — Instantaneous field of view
Pixel — Picture element
ERE — Effective resolution element

Regarding Image Interpretation the process is the same with the aerial photographs when the visual interpretation is made i.e.

1. Recognition & Identification
2. Analysis
3. Deduction
4. Idealization and finalisation

Besides these there are Spectral, Radiometric and Temporal resolution. These are required for different purposes and different experts have analysed the same.

(v) The centre should have the capability of handling remote sensing data from variety of platforms.

(vi) The centre should have the facilities for efficient data storage and retrieval.

(vii) The centre should have the capacity for making the overview analysis on a national or regional scale.

Such transfer of technology is both good and bad. It should be used judiciously for betterment of human beings. So it can be said that it is a double edged sword with political and legal implications.

Resolution is an important thing for a photo-geographer. It is more so in this work, as it is completely based on aerial photographs and satellite imagery interpretation. Real meaning of resolution is finer by which an object could be identified properly. But here we generally refer to as spatial resolution which may be defined as the minimum distance between two objects that a sensor can record distinctly. We may call it the overall fitness of the detailed character of the image. It has been suggested by Forshaw and others (1980) that the definition of spatial resolution be placed into one of the four categories, viz. ,

(i) The geometric properties of the imaging system;
(ii) Ability to distinguish between point targets;
(iii) Ability to measure the periodicity of repetitive targets; and
(iv) Ability to measure the spectral properties of a small finite objects.

Image Analysis Paradigm

Image analysis is a super task over aerial photographs. Years of work by some far sighted innovative professionals have laid the foundation upon which analysis of remotely sensed data is based. Most recent work and advances in the field of remote sensing have resolved around non-photographic imaging and computer assisted processing techniques. There is need to critically re-examine the basic elements and principles of image analysis which are common both in manual and computer-assisted techniques. Thus, we find that an artificial barrier has been erected between visual interpretation and computer assisted technology or automated image analysis. But these barriers are false and have only made it more difficult to discern either fundamental similarities or difference between two approaches. There are good deal of discussions about these for image interpretation. Now 'general agreement' stands on the following paradigms. (i) Shape, (ii) Shadow, (iii) Tone, (iv) Colour, (v) Texture, and (vi) Pattern.

The photo-key is of utmost importance to the photo-geographer. A little clarification is needed here. When many objects are imaged, interpreter must be aware of factors not necessarily related to their experience or to the work of their organisations. This means familiarity with so wide a variety of stimuli that even the most accomplished interpreter is occasionally dependent upon the reference materials. Identification kit used with great success by field workers in physical sciences particularly in the subject of Geology and Physical Geography are the useful aids of interpretation. So the photo-key should be oriented as per the desire of the interpreter itself. Photo image includes all the elements which are on the surface, but the choice of the interpreter is selective.

3

GEOGRAPHICAL BASE OF THE SELECTED AREAS

For the application of Remote Sensing Technique and aerial photo-interpretation, three areas are chosen on the basis of geographical factors like physical setting, relief and landform etc.

(i) The first area selected is in the Dun Valley, confined between the Lesser Himalaya and the Siwaliks. The Dun Valley is famous as it itself forms a morphological unit called Duns. Many Duns of the Sub-Himalayas are examples of tectonic valleys, these being synclinal troughs enclosed between two contiguous anticlinal flexures. This aspect is however, somewhat modified by deep alluvium and sands which have filled up the bottoms as well as the slopes of the bordering mountains.

(ii) The second area selected is a typical plateau region. It is in Sambalpur district, Orissa. Here peninsular formation is predominant. It is an undulating upland, varying in elevation from 150 to 225 m above sea level excluding hills and table lands. The general slope of the land is from north to south and the consequent drainage system is merged with the Mahanadi river and its tributaries. But the area is selected because of a significant man-made feature which has been established here in recent years. This is a major multipurpose river valley project, the first of its kind in this State. This river valley project has led to the laying of a network which is commonly known as Hirakud Canal Command Area, irrigating about 267,448 ha of land, thereby enabling to produce two to three crops of rice in recent years. As a result, the whole scenario of the area is completely changed from dusty brown to lush green. The basic reason of selection of this area is to monitor any change in the socio-economic condition of the area due to Hirakud canals. Formerly the people of this area used to suffer from acute financial constraints and poverty. Now the condition has been completely changed. The people of this area have become wealthy and prosperous. Some new towns have grown up in the area. A new University, Medical College and Engineering College have come up. New industries are developed. Aluminium and cement industry bear testimony of industrial growth of this region. Landuse mapping has been carried out in this area with the help of aerial photographs and satellite imagery. About 100 km. north-west from the town Burla and 50 km. from Bargarh town, limestone is being quarried at Dungri for Hira Cement Factory, which is located at Bardol near Bargarh town. This cement factory is producing one of the finest quality of cement

(iii) The third area selected is the Mahanadi delta alongwith he coastal plain of the Bay of Bengal. The selection of the area is mainly to monitor any change that has taken place here as a result of the Hirakud Dam on the Mahanadi river in the upper course. The Mahanadi literally means the great river and is one of the biggest rivers of India flowing through the States of Madhya Pradesh and Orissa. In this process devastating floods frequently used to occurr in the lower deltaic plains in the coastal districts of Puri and Cuttack. The flood damages were so devastating that economic back-bone of the people of these two populous districts were shattered completely. Similarly the lands in the District of Sambalpur in the Western Orissa were facing often drought situation due to erratic rainfall pattern and layout of land even 'though this mighty river was flowing through them. Now floods in these two districts have become rare phenomena. Hirakud reservoir supplies irrigation water through the river as and when required. Further it has been observed that due to construction of the mighty Hirakud Dam on the Mahanadi, the siltation process in the mouth of the delta i.e. in the Bay of Bengal has considerably been decreased. As a result the delta building activities of this river have been considerably weakened. This area throws some light on the good and evil effects of the Dam construction. Analysis of the three conspiquous regions of India depict the role of aerial photo and image interpretation for different purposes and not for a solitary one. These tools help immensely for indepth studies of terrain, soils, social and economic attributes of the various regions.

The distinct geographical personalities of these areas are noted below :

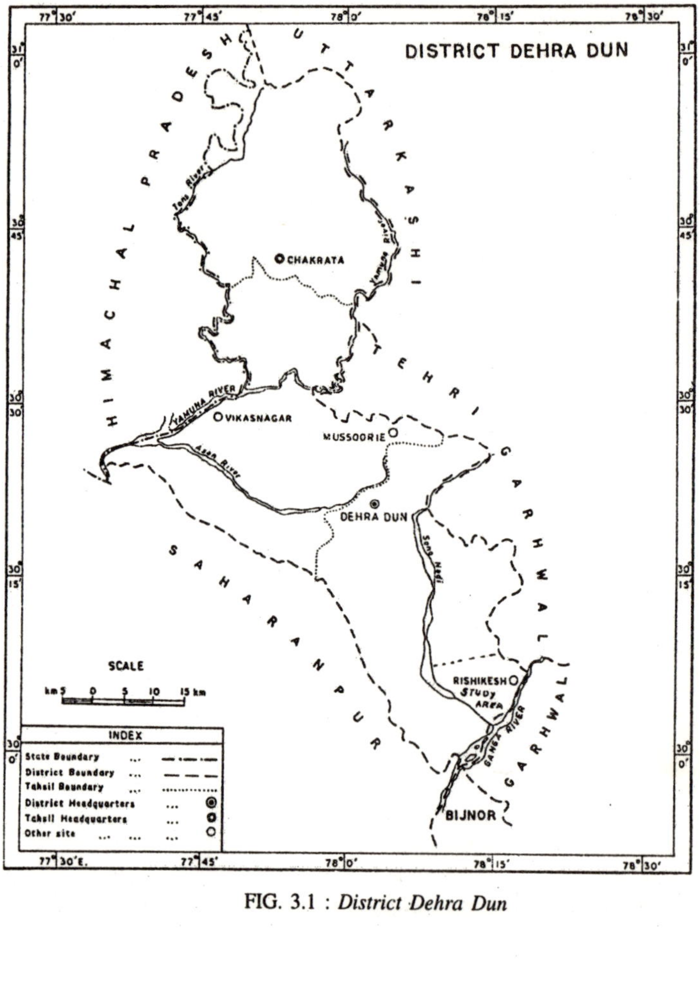

FIG. 3.1 : *District Dehra Dun*

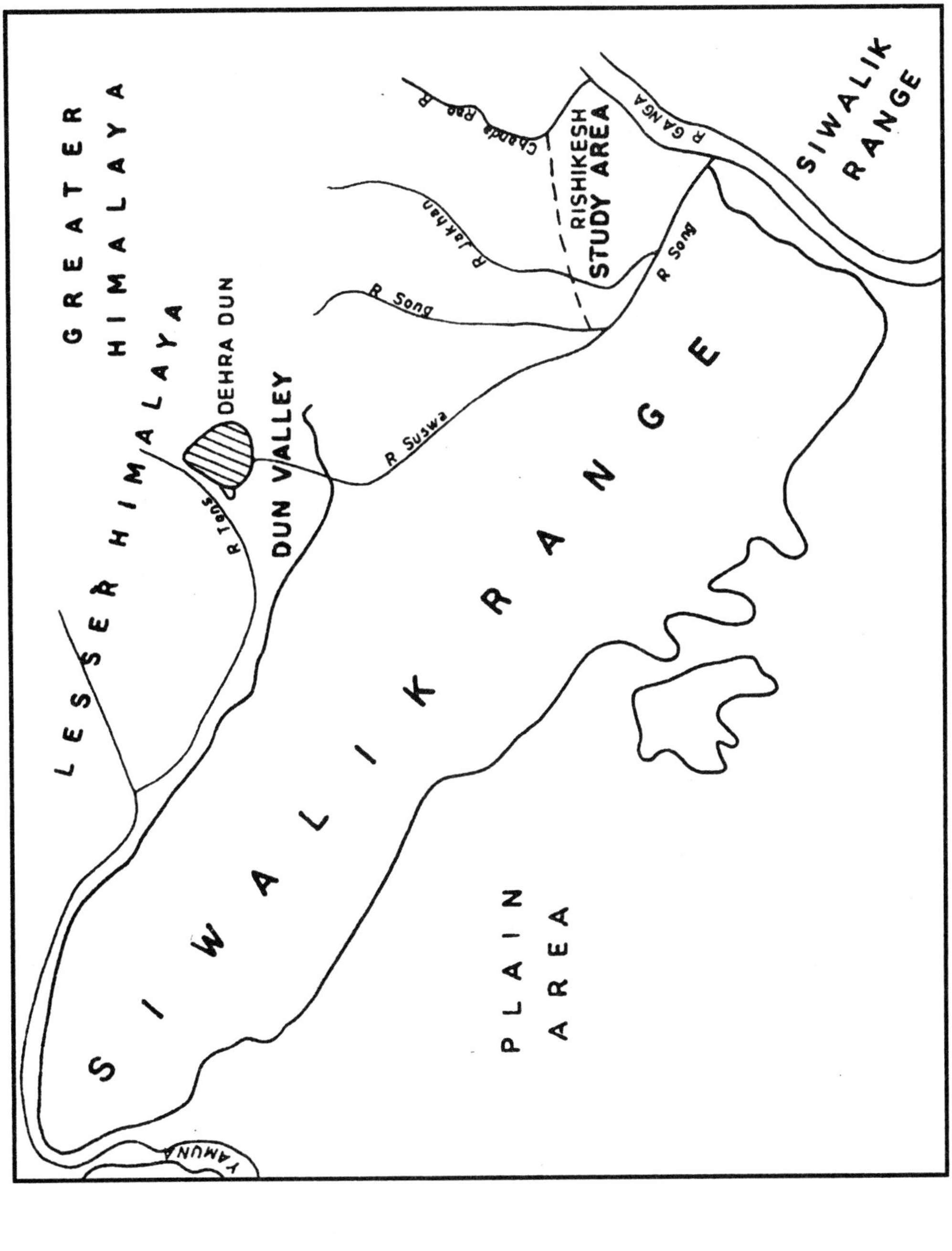

FIG. 3.2 : *Satellite Imagery Showing Study Area Enlarged of Dehra Dun and Rishikesh* *(Contd. on page 48)*

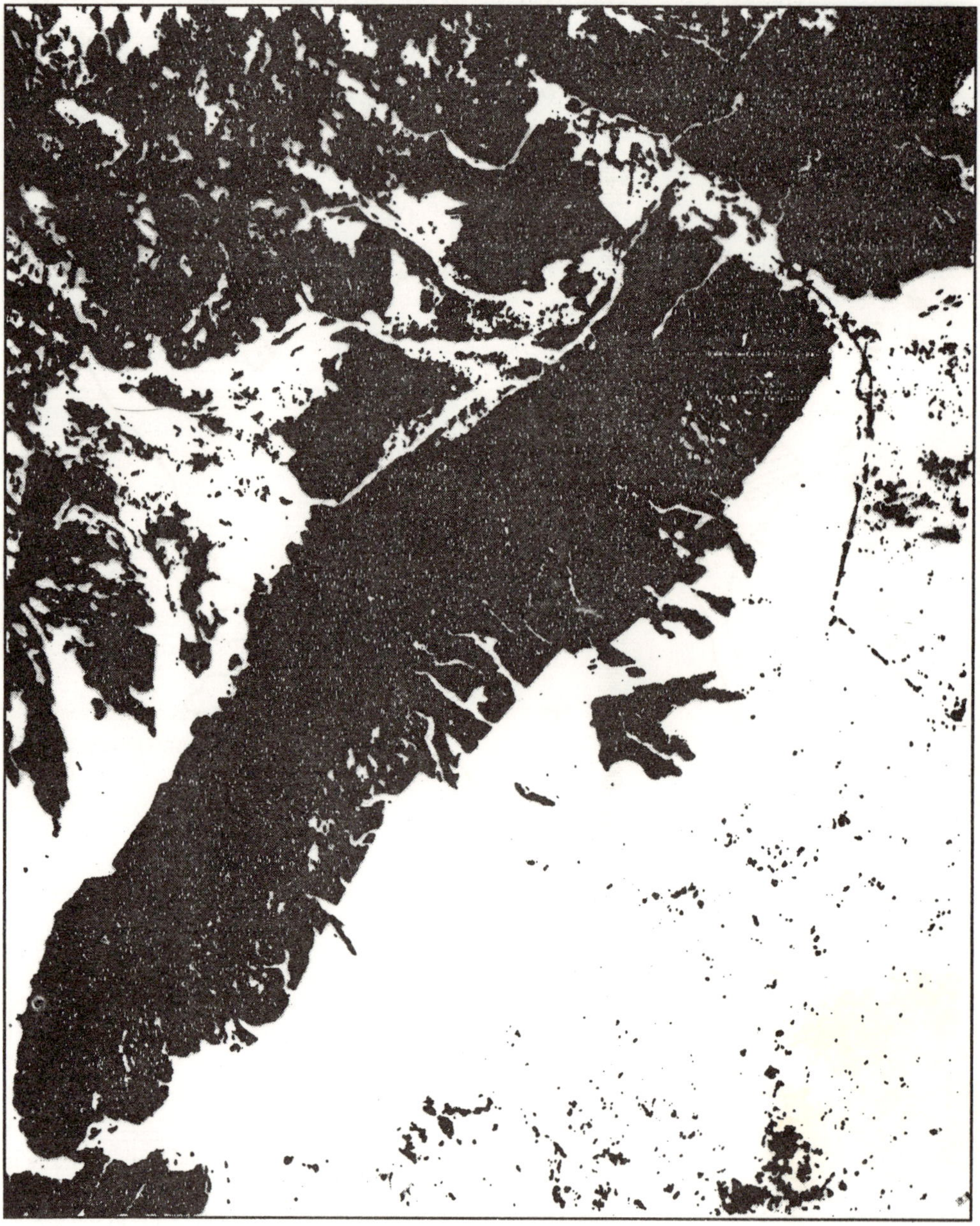

FIG. 3.2 : *Satellite Imagery of Dehra Dun and Rishikesh*

The Himalayan Foothills

This region lies in the District of Dehra Dun. The administrative division is very peculiar unlike other Districts of Uttar Prasesh. The number of Sub divisions exceeds that of tahsils. There are three sub-divisions (a) Chakrata, (2) Dehra Dun, (3) Mussoorie. Dehra Dun is located on the Dun valley where as Mussoorie is on a mountainous area. Chakrata, the northern tahsil and a sub division of the district is formed by the hill pargana of Jaunsar Bawar and is roughly elliptical mass of mountain stretching due north from the Lesser Himalaya .

The southern half of tahsil is encircled by the Tons river in the west and the Yamuna river in the east and south. Chakrata is bounded on the west by Himachal Pradesh and east by Uttarkashi. Dehra Dun sub-division is bounded by Garhwal and Tehri Garhwal on the east and on the west by Saharanpur district and the southern side is touched by Bijnor District. The photo-geography is, studied mainly between the Ganga-Yamuna Doab.

The Siwalik hills of moderate height is located between these two mighty rivers. The pebbles are of different size and are completely rounded and this suggests that pebbles are rounded due to splitting, cracking alongwith velocity of the stream water and they are carried from long distance. It has also been observed during field work that the size of the boulders are increasing as digging is made down-ward. In the Dun valley the well is dug up to 33 meters depth to get potable water from the well. This shows that the groundwater level at or below 33m from the surface. The Dun Valley, once upon a time, was covered with grassy swamps. Tall grasses grow in abundance is this Valley, because nearly half of the valley belongs to the tarai region. It has been estimated 350m is limit of the *tarai* and *bhabar* region. So below 350m contour, most of the land is within the *tarai*, being infested with tall grasses and wild animals. In recent years much of the original grass cover have been cleared to make room for farming. The land is most suited for the cultivation of sugarcane, being followed by wheat and gram cultivation.

The district comprises of two distinct tracts, viz., the mountainous tract which include the Chakrata tahsil and the Mussoorie Sub-division of Dehra Dun Tahsil and sub-mountainous tract is the Dun valley .

Mountainous Tract

This part is complicated and many parts are not accessible. In this case Satellite Imagery and aerial photographs are of immense help. Imagery and aerial photographs give clear pictures and ideas to decipher relief and structure of the land and open new vistas for further exploration. These tools are also helpful in drawing detailed maps. Firstly the detailed study is carried in Dehra Dun District with the help of aerial photographs and subsequently with the Landsat Imagery of band 5 and 7.

Sub-mountainous Tract

Below this mountainous tract lies the Dun proper which is an open valley bounded by the Siwalik hills in the south and the outer scrap of the Himalaya. The Dun is irregular parallelogram in shape with its longer axis lying almost due north-west and south-east and its existence as such is due to the presence of Siwalik hills, as well as marked range extending along the entire south western border. To the south they present a steep face scarred by frequent precipices. On the north a long gentle flank slopes inward and forms a longitudinal shallow valley terminated by the outer range of the Himalaya. The drainage of the valley is borne by two rivers rising at no great elevation, a few kilometres south of Dehra and

FIG. 3.3 : *Satellite Imagery Band 5 showing Dun Valley in white patches in the north of the Siwalik Range, (Scale : 1:3,000,000)*

flowing east in the Ganga river and west into the Yamuna river. Both the rivers Ganga and Yamuna have come down on the plains through the break in the Siwalik. The Dun valley is apparently a single valley but in reality it belongs to two great river systems, those of the Ganga and the Yamuna river. The Dun is a well wooded plain and its surface is diversified. There are two small hills known as Nagsidh and Kalanga both near Dehra Dun, which have broken the monotony of the plain.

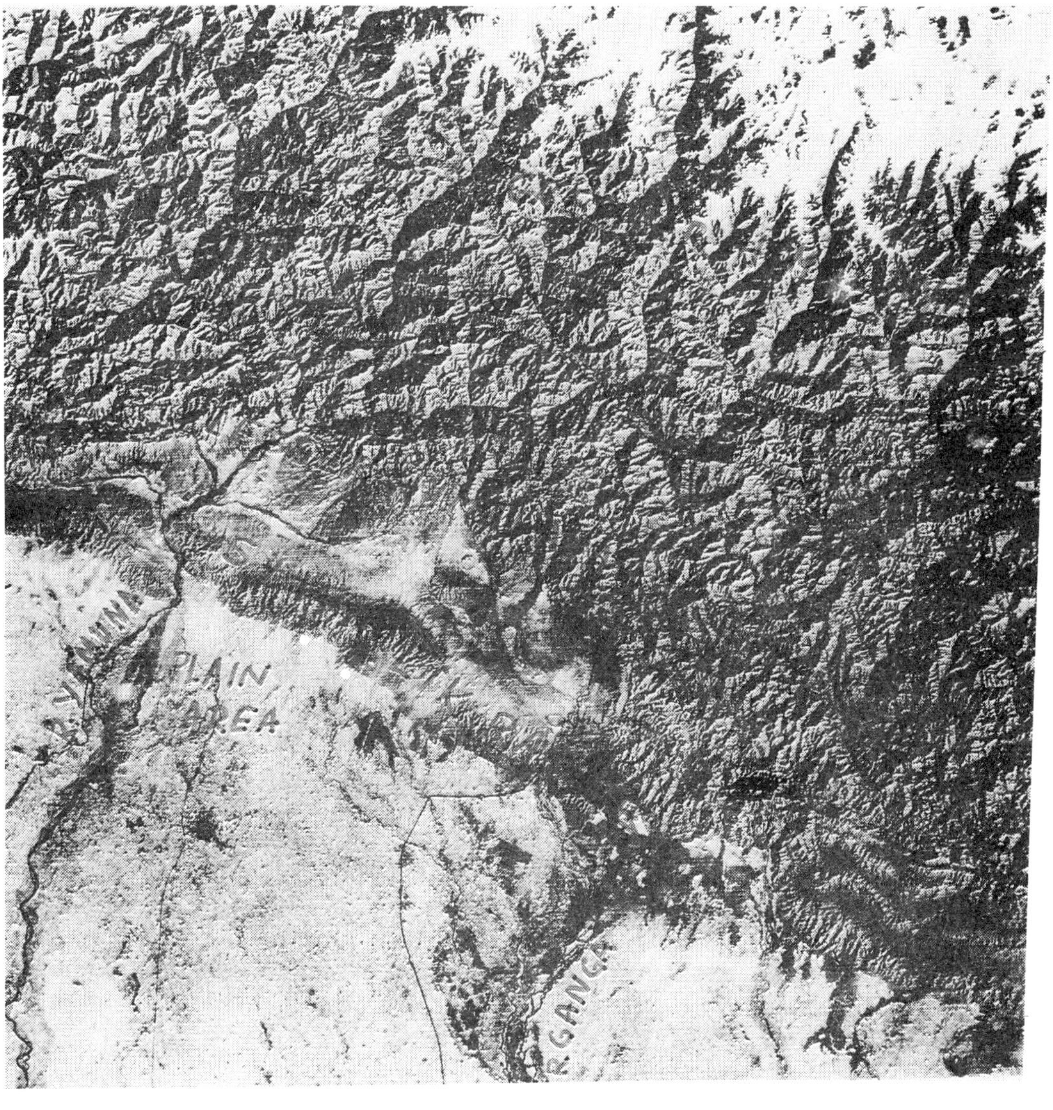

FIG. 3.4 : *Satellite Imagery Band 7 showing Siwalik Range, Lesser Himalaya and Greater Himalaya with Permafrost area, (Scale : 1:3,000,000)*

Hirakud Canal Command Area

The Hirakud Canal Command Area is developed after the Hirakud Dam came into existence. It is one of the first major multipurpose river valley projects of India. It has the distinction of being longest dam in the world being 25.8 km. long (Dam and dykes taken together) . It has also the rare distinction of forming the biggest artificial lake in Asia with a reservoir spread of over 737 sq.km. The project provides direct irrigation from the Dam to 159,063 ha of kharif crop and 108, 385 ha of rabi crop totalling to an irrigation spread of 267,448 ha of land of this State. The water released through the

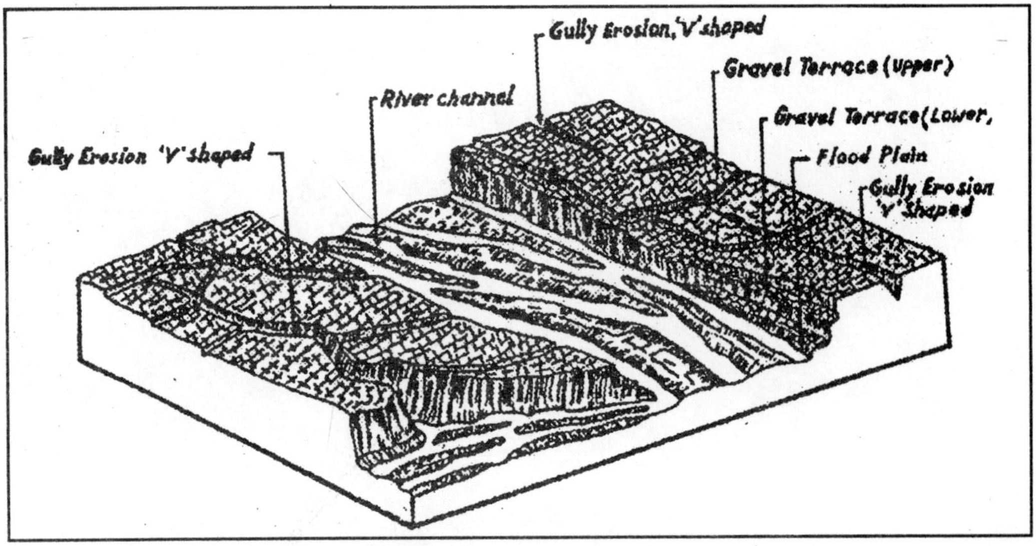

FIG. 3.5 : *Typical Block diagram of Air Photo Pattern for Gravel Deposits Dun Valley*

Dam is of the order of 25, 000 cusecs, irrigating 436, 000 ha in the Mahanadi delta in the districts of Cuttack and Puri in addition to stabilizing irrigation of 78, 540 ha in the two districts.

Sambalpur district consists of a wide expanse of fair open country fringed by forest clad hills as well as series of low hill ranges of extremely irregular shape. The district is an undulating upland varying in elevation from 150m to 225m above the sea level excluding hills and table land, the general slope is from north to south. But it is much broken up by rugged ranges and is traversed in all directions by drainage channels mostly leading from the hills ranges to the Mahanadi river and its tributaries. Isolated hills rising abruptly from the plains and hill ranges are also common and a considerable area consists of ground cut up by ravines or broad sandy ridges. The Mahanadi river enters the district in the north-west of Sambalpur sub-division and flows into the Hirakud reservoir which covers an area of 774.41 sq.km. when the storage water stands at 192 meters. After crossing the Hirakud Dam it takes a wide curve and moves to the south into Sonepur sub division of Bolangir district. The principal tributary of the Mahanadi is the Ib which enters Sambalpur from Sundergarh district in the north.

The following geological succession may be recorded :

Recent
Tertiary
Lower Gondwana
Cuddapah
Dharwars
Archaeans

The Archaean metamorphics are probably of Dharwarian age. Around Sambalpur, Burla and Hirakud, the strike is variable in places becoming ENE - WSW .

The principal crops of the district and Hirakud Canal Command Area are paddy, pulses, oil seeds, sugarcane, millets and wheat. The total cropped area of the district is about 601,425 ha of which paddy is the staple crop occupying about 84 per cent of the total cropped area. The method of cultivation has changed in Canal Command Area. In the irrigated area transplantation is predominant

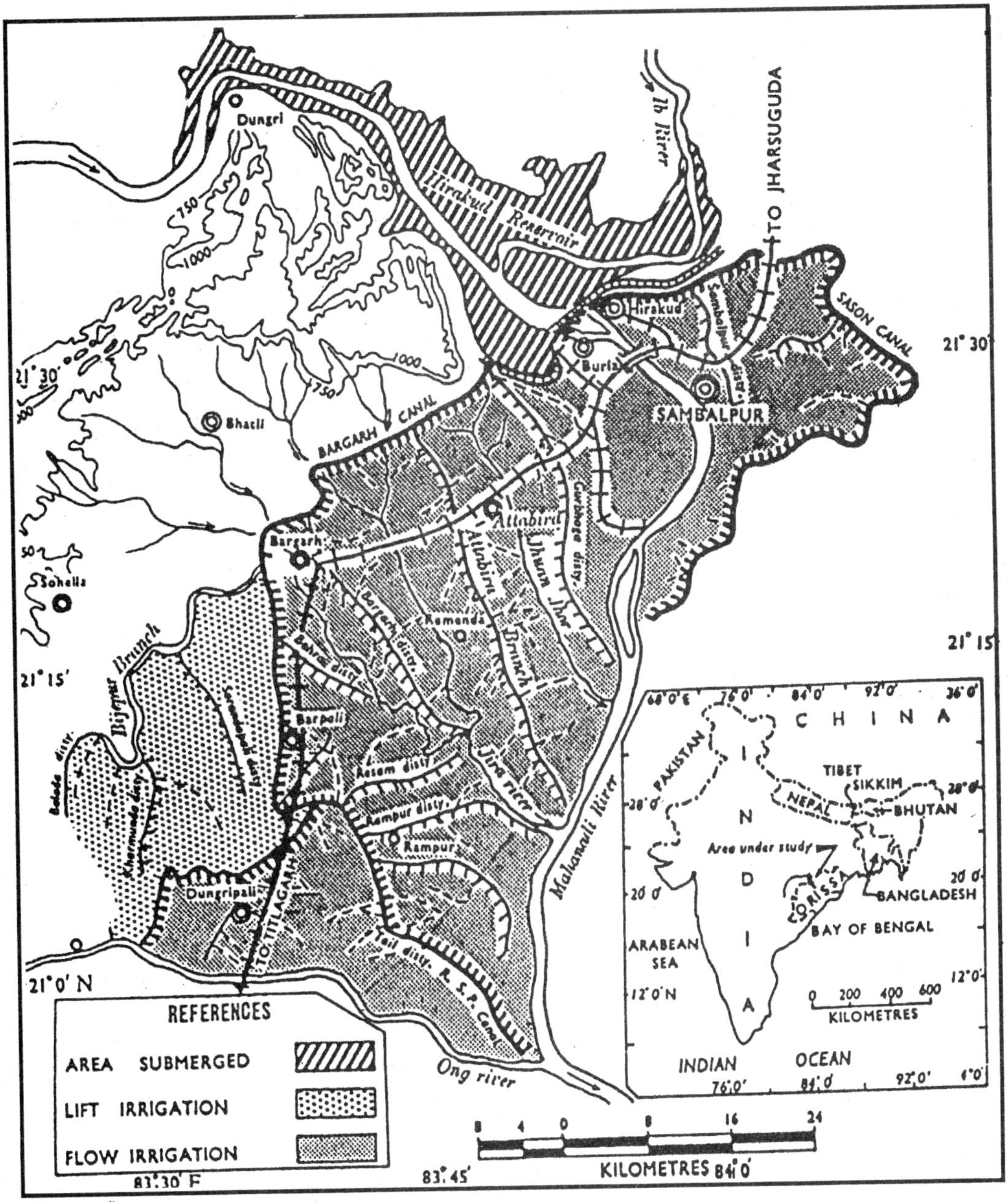

FIG. 3.6 : *A Key to the Hirakud Canal Command Area, Orissa.*

over broad-cast method, as it is more remunerative. But in the non-irrigated area broadcast method of sowing is still widely practised.

According to local relief the land is classified into four categories locally known as (1) At, (2) Mal, (3) Berna, and (4) Bahal. The 'At' land is at higher elevation than the 'Mal' land and 'Berna' is at lower level than 'Mal'. The lowest flat land is called 'Bahal' and different crops are cultivated

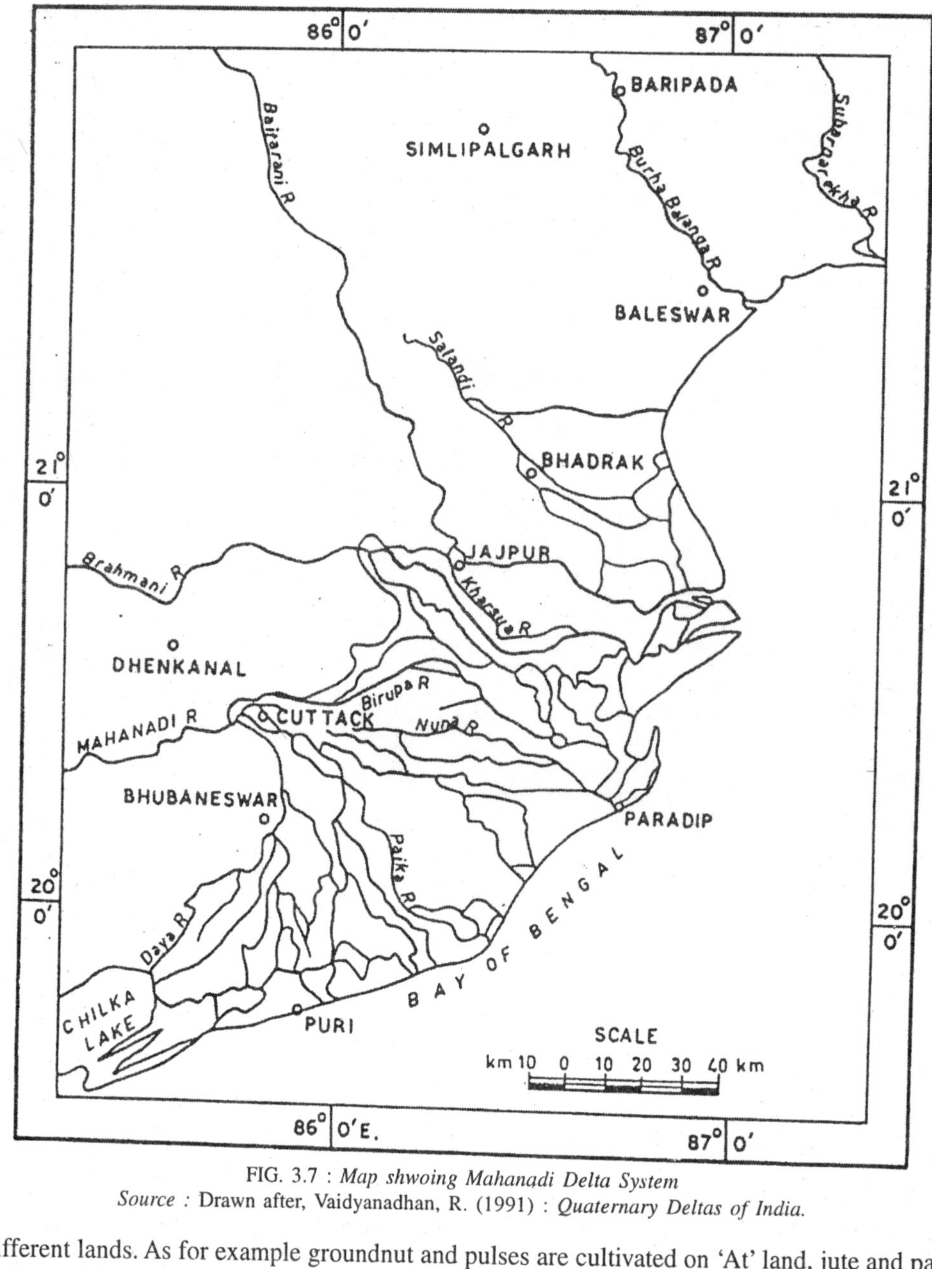

FIG. 3.7 : *Map shwoing Mahanadi Delta System*
Source : Drawn after, Vaidyanadhan, R. (1991) : *Quaternary Deltas of India.*

on different lands. As for example groundnut and pulses are cultivated on 'At' land, jute and paddy are cultivated in ' Mal' land . Paddy prefers 'Berna' and 'Bahal' land . So the overall study reveals that cultivation is the dominant economic activity of the people. Some however depend on forest, such as on timber products, sal seed.

Coastal Plain of Puri and Cuttack Districts

It is difficult to demarcate the Mahanadi Delta because of the over lapping of the deltas of the

Mahanadi with the Brahmani. The Mahanadi river in combination with the Brahmani and the Baitarani have been able to form an extensive alluvial tract extending from Balasore in the north to Chilka lake in the south. The delta is about 175 km long 75 km wide. A distinct feature of this delta is that it has got its maximum growth not in the centre but in the northern part. Prior to the formation of delta the original shore line was concave towards the sea owing to the hard bed rocks of the Eastern Ghats the delta falls into two broad sub-division.

1. The old delta of the older alluvium of the tertiary period.
2. The active delta of the newer alluvium of recent period.

The land-form map of the delta prepared from satellite imagery MSS Bands of 4, 5 and 7 has shown the region as a flood plain of newer alluvium and moderately high land may be marked as the old delta of Tertiary period. The red coloured patches are the hillocks which are away from the deltaic portion. These hills are of Dharwarian age. There are frequent patches of laterites. Landuse study of this region has also been carried out. Sketch maps of landuse and land-form based on Imagery are also presented and a tentative classifications of landuse and land-forms have been worked out.

4

PHOTOGEOLOGY OF DEHRADUN AND ITS NEIGHBOURHOODS

Introduction

Satellite Imagery and aerial photographs for geological and geographical mapping particularly of geomorphology play significant and important role. From satellite frames and detailed studies of aerial photographs, the regional structure and generalised stratigraphy can be interpreted without carrying out extensive field work. In some cases, thorough ground checking could not be carried out, because of inaccessible terrain but with the help of Landsat and T. M. data sufficient information can be gathered to overcome these difficulties.

The structure, stratigraphy and correlation of different formation of U.P. Himalaya and Himachal Pradesh have been considered here. Himalaya have been described differently by various workers. At present some idea could be generalised with the modern technique. With Satellite Imagery of the scale of 1:1 million supplemented by the aerial photographs, perspective maps can be drawn to focus various morphological phenomena.

Previous works of different geologists are very helpful for collateral information to study the geomorphology and geology of any area. One of the pioneers to study this terrain was Oldham (1883, 1885 & 1888) who gave first a detailed account of the geology of Jaunsar-Bawar and Chakrata area and made passing reference to the Deoban rocks north of the Tons river. Pilgrim and West (1928) mapped Shimla and Chaur region and applied the principle of low angle thrusting.

The complex stratigraphical feature of the Lesser Himalaya of Uttar Pradesh and Himachal Pradesh can be successfully tackled only if systematic mapping can be carried out on the basis of the Satellite Imagery. Much has been achieved with the help of photographs during this work. Otherwise entire stratigraphy and correlation of this part of Lesser Himalaya have been solely based upon the *grade of metamorphosis and lithological variations* which have been affected by *large scale thrustings,* rendering such correlation *dubious in nature*. The *calcareous formation of Uttar Pradesh near Mussoorie* and *Himachal Pradesh Himalayas overlie on association of quartzite, phyllites,* greywack slates and basic flows which from the floor of Deoban and Shali sedimentation. Younger arenaceous group (Sandra & Shali quartzite) make the last phase of Deoban and Shali sedimentation. The upper quartzites and slates of Deoban formation continue as a *para-autochthonous* unit below the *Krols,* in Chandpur area. The entire sequence of *Krols* occupies higher horizons in stratigraphy and can not be correlated with Shalis or Deobans as suggested by West (1939) . The thick agrillaceous, arnaceous and calcareous formation (Deoban Shalis, and Chandpuras) all along the lower Himalaya

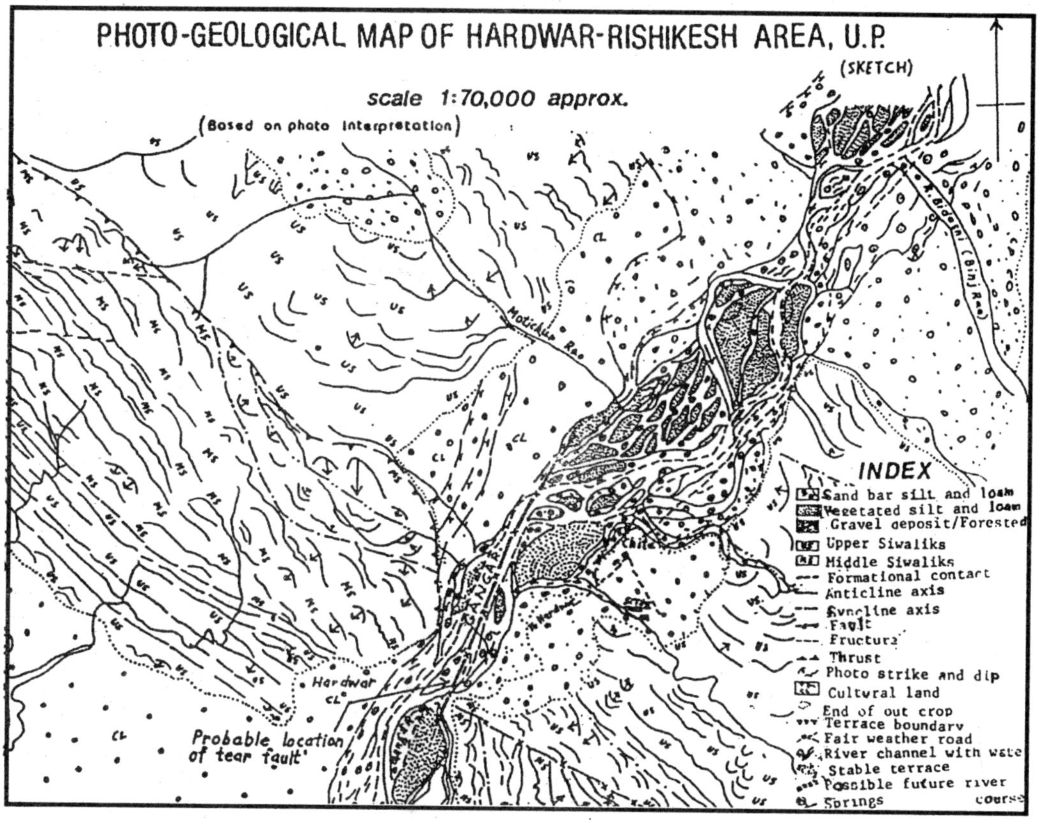

FIG. 4.1 : *Photogeology Map of Hardwar and Rishikesh Area*

Source : Based on R. P. Sharma IPI, Appreciation Seminar 1972

may be of *marine origin equivalent of the Peninsular Vindhyas.* The recent discoveries of a stromatolitic horizon in lower Himalaya section support to this analogy. It suggests a late Precambrian to Cambrian age. The limestone turns suprisingly thick section reaching several thousand metres in the Sutlej, Tons and Yamuna valleys and they are overlain by equally thick quartzite section in the Bhagirathi and Alakananda Valley assuming that the entire sequence is normal.

The different members of Deoban formation exhibits *tremendous facies* variation along and across the strike and this presents great difficulty in delineating the various units. Sandra quartzites which is the major unit in the east laterally grades into shales and phyllites in the west. The underlying Tikri slates also show similar behaviour. The Isu Tibba limestone with thick section forms the landmark in the west but when traced towards Tiuni it grades into slates, phyllites and carbonaceous shales with only a few lenticles of limestone. Thus, the facies change in the Deoban formation presents further difficulties in correlation on regional basis. It can only be established by strike and lineation marking on the aerial photographs, and landsat imageries.

Structure

The rocks of Shali formation occur in two structural belts. The main belt forms two bifurcating

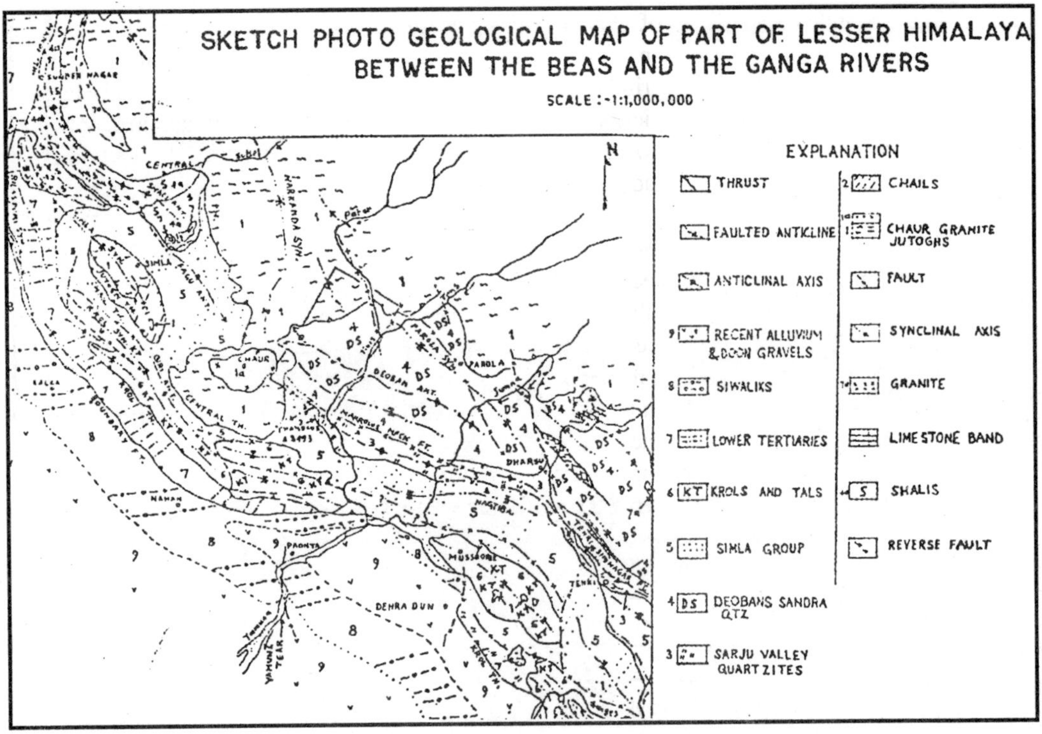

FIG. 4.1 (a) : *Photo-geological Map of Part of Lesser Himalaya between the Beas and the Ganga River*

Source : Based on R. P. Sharma, Appreciation Seminar, IPI DDN, 1972.

digitations-one terminating at Tattapani (31°14' 50" 77° 0510") and the other along the ridge of Nag Tikar in the type Shali area. The Sutlej valley syncline separate the main Shali Anticline and Tattapani anticline. The main Shali anticline is further faulted by a steep angle reverse fault in the Chaba-Batwara area. The Tattapani digitation presents complex assemblage of faulted anticlines and normal well preserved synclines. The main Shali belt beyond Alsindi is complexly folded and faulted with the result older units are seen to ride over the younger rocks, in Gairu Sundernagar area. The subsidiary structural belt apparently independent of the Main Belt, extends from Bahind Tibba to Khadli with its best development in the Bilaspur area over a length of 56 km almost north-south direction. Since this digitation constitutes the outer limit of para autochthon, zone, it has been subjected to complex folding and reverse faulting. As such it is asymmetric syncline with both limbs faulted.

The annular drainage system of the Sutlej, Nauti and Giri rivers conform with the regional anticline structure of the Shalis. The *shells, slates, plyllites* and *quartzites* (Chadis of West 1939) overlying the Shalis in Nauti Khad and Tattapani area belong to the rocks of *Shimla Group* which is a younger stratigraphic unit lying over Shalis. This rules out the possibilities of Shali thrust in the *Nuntikhad*, and Tattapani area suggesting that Shalis do not form a techtonic window as described by West (1939).

The Deoban formation also occurs as faulted ' Semidom' at the closure in the Chaupal Peontra area and extends with tremendous *facies variation* towards Jamuna and the Bhagirathi valley. In the north it is bounded by the *central crystallime thrust sheet,* which occurs as youngest tectonic unit in the area. The Southern boundary of Deoban formation is defined by a regional faulled anticline which extends from the Tons river in the North and to the Alaknanda river in the South. The oldest autochthonous rocks gradually described as Morar Chakrata bed, Shimla slate *Sauri* Gad Slates are exposed along the trace of this anticline. Other structures of regional importance in Deoban are Deoban syncline, *Parola-Dharasu syncline,* Naugaon syncline and Uttar Kashi, Tehri anticline. The east-west trending faults between Rudraprayag and Karnaprayag control to course of the Alaknanda river.

The rocks of Deoban formation are intricately folded and fallted under observation. The Southern boundary of Deoban belt it is marked by a steep angle reverse fault which brings Tikri Slates and Sandra quartzite in direct contact with Morar-Chakrata beds (Shimla slates) of Oldham (1988). The same fault in the east has been mapped as the Merrows neck fault. In the north and east the Deoban belt is overlapped by crystaline and metamorphics of Chails and Jutoghs.

The various members of Deoban formation traced from Tiuni in the north continue further west and south with a regional swing in the strike from nearly east-west to north-south in the strike from nearly east-west to north-south and finally north west-south and east in the Minas river. This brings out a semidomal pattern with a northwestern plunge. The crystallines and metamorphics (Chail & Jutoghs)

Table 4.1 : Geological Sequence of Dehradun Area

		Sharma (1972)	*West (1939)*	*ShriKanta & Sharma (1969)*
		1	2	3
		Basic Sills and Dykes		
		Malat slates : Grey & brown slates with minor quartzite	Madhan slates : Rusty brown sandy slates slightly micaceous	*Bandla Member :* Shales silt-stone quartzitic breccia
		Sandra Member : Ortho-quartzites with minor slates and basic rocks	Shali : Quartzite : Pure white quartzite sometime containing chert	
		Tikri Member : Slates with minor bedded limestone and quartzite	Upper Shali Limestone : Massive grey, dolo-mitic limestone with only occasional chert	*Paranli Member :* Cherty dolomite, grey limestone and white quartzite
DEOBAN	UPPER DEOBAN			
		Isu Tibba Member : Massive to bedded limestone with minor shales and slates		
	MIDDLE DEOBAN	Peontra Member : Slates, quartz and conglo-merates	Shali slates : A variety of slates and slaty limestone	*Makri Member :* Grey, black and purple shales and slates with thin bedded limestone and quartzite — also spora-dic cherty dolomite

(Contd.)

(Contd.)

		1	2	3
FORMATION	LOWER DEOBAN	Meenas Member : Stromato-llitic limestone and black shales	Lower Shali Limestone : 2. Massive, grey dolomitic limestone, full of parallel sheets of chert	*Tattapani Member :* Cherty dolomite, platy limestone and shale
			1. Pink calcitic limestone with no chert, banded at the base	*Sorgharwari Member :* Pink & grey limestone with sporadic shale partings
				Khatpul Member : Massive dolomite with a thin red shale horizon at the base
		Atal Quartzites : Pink and white quarzite with slates and basic rocks	Haira Quartzite : White and purple quartzite	*Khaira Member :* Purple and at places with quartzites
				Ropri Member : Brick red shale and silt-stone with an impersistent band of dolomite and salt bed at places
				Mandi Volcanics : Volcanic flows intrusives
				Sundernagar formation: 2. White & purple quartzite with shale; 1. Shale, slates with quartzite bands

Source : West, W. D., Srikanta, S.V. *et al.* (1969) : *Bull. Geol. Sur. Ind.* Vol. 6,3

occupy a higher tectonic position and form the true *allochthon over Deoban.* The impact of the thrust sheets on the Deoban autochthon is so pronounced that the most of the folds in Deoban formation show westerly and south westerly overturning and the consequent over thrust of the older Deoban members over the younger. This has presented a typical riding of the blocks over another. The area connecting the Isu Tibba with Tons, from the main antiformal complex with domal wraps and overturned folds. There are two styles of folds one along NW-SE direction and another nearly west-east with a westerly plunge. In the later type is an isolated and a complex one and is seen along a thrust front. Isu Tibba syncline lies south west of Atal anticline and north of Minas anticline and is an important and complex structural feature. The core of syncline consists of quartzite and slates. The Atal Minas and Pabar anticlines are the most important structural feature in the area. Table 4.1 showing morphology are attached.

Morphological Features

According to West (1939) , the following morphological characteristics are observed :

(1) *Madhan Slate :* Rusty brown sandy slates slightly micaceous.
(2) *Shali Quartzite :* Pure white quartzite sometime containing chert.

(3) *Upper Shali Limestone :* Massive grey, dolomitic limestone with only occasional chert.
(4) *Shali Slates :* A variety of slates and slaty limestone
(5) *Lower Shali Limestone:* Massive, grey dolomitic limestone, full of parallel sheets of chert. Pink calcitic limestone with no cherts banded at the base.
(6) *Khaira Quartzite :* White and purple quartzite

A comparative chart of Sharma (1972), West (1939) and Srikanta and Sharma (1969) is appended here for full understanding the morphology of the area.

In conclusion it may be said that the detail study of the structure is only possible due to aerial photographs and land set Imagery particularly of band 5 and 7. The air photographs and imageries unveil different structural units and it is easy to delineate the boundaries in detail with a field check. Distant places are recognised by these tools and extrapolated the ideas by extending the possible alignment, discontinues and collateral information of the previous works.

5

GEOMORPHIC STUDY OF RISHIKESH AREA

The area surveyed is in the south-western part of Dehra Dun district and is bounded by the river Song on the west and the Chandna Rao on the east. The south-eastern portion is bounded by the mighty river Ganga (Mazumder and Banerjee, 1967) . In a broad sense it is a small tract south of the Nagasidh, Most of the part is occupied by the reserved forest in the Song-Jakhan interfluve, a small agricultural land called Chidrawala, an important locality for settlement. Otherwise the forest is devoid of any settlement. The area lies between 78°10' E to 78°20' E longitude and 30°2N to 30°7N latitude and is approachable from Dehra Dun by Dehra Dun Rishikesh metalled road and unmetalled forest roads.

Climate

As the altitude of this region ranges between 330 m to 600 m, the climate is more temperate than that of the adjoining plains. The monsoon climate in the Dun has three seasons : Winter, Summer and Rainy. The winter begins at the end of September and prevails up to the middle of March. The temperature falls nearly to freezing point by the end of December or early January. Frost is common in low-lying tract in winter nights during December and February. This causes damage to young plants to a great extent. During Summer that is March and April, the temperature rises rapidly, though this rise is moderated by occasional thunderstorms. The temperature continues to increase during the month of May and first half of June, though the night often remains cool as a result of a cool breeze blowing down the hills. But sometimes the nights are hot, particularly during the last days before the commencement of the rainy season. The maximum daily day temperature varies from about 20°C in January to about 41°C in May and June. The monsoon generally sets in by the middle of June and prevails until about middle of September. In July and August the rainfall is heaviest. Rain does not generally occur between October and December. Winter rains, which are sometimes heavy and continue consecutively for two or three days, fall mainly in January or early February. There after there may be no rainfall until the commencement of the monsoon. However, occasionally short and sudden showers accompanied by thunderstorms may occur from March to May.

Vegetation

As per Champion's system of classification the area comes under North Indian subtropical moist deciduous forest type reveals the climax of vegetation types. The area possesses a large treasure of flora. In some places reckless cutting results in creation of bad lands. The main timber species is the Sal (*Shorea robusta*).

Geology

There are three main geological formations in and around the Dehra Dun Valley. These are –

(i) Dun sediment
(ii) Siwalik Rocks
(iii) Himalayan formation

The area under study covers the *Dun alluvium*. The Dun Sediments constitute the *Quarternary fluvial filling* which form a synclinal *Valley in the Siwalik formations*. The alluvial bed is of irregular thickness and covers practically the entire Dun. It is formed of vast accumulation of clay, sand, silt, boulders, pebbles and gravels carried by the Chandna Rao, Jakhan Rao, Song, Ganga and tributaries to these rivers.

During formation of the Himalaya the rocks have been folded, faulted and even carried to considerable distances due to overthrusting. They are meta-sedimentary rocks embedded within igneous intrusion (mainly granitic) . Prior to the tectonic disturbance of the Tertiary era, the region belonged to the part of the vast Tethys Sea. At a later stage during the late Tertiary era and the early Pleistocene, the Siwalik system of hills, which form foot-hill zone of the Himalayas were formed. The presence of mammalian fossils is one of the characteristics of the *Siwalik rocks* which consist of boulder beds of coarse sandstone and clay .

There is a synclinal valley locally called Dun between Siwalik range and the Himalaya. This valley comprises a geological formation of *Dun Gravels* which were deposited during Pleistocene and Recent period. There is sub-division of the Dun alluvium into two hydrographical formation.

(1) *Bhabar* (2) *Tarai*. The Bhabar tract is one with a higher elevation (boulders and shingles) through which water sinks readily and disappears. The *Tarai* occurs at a lower elevation. In *Tarai* the water that seeped into *Bhabar* emerges and forms numerous temporary or sometimes perennial meandering streams .

Interpretation of Aerial Photographs

Before systematic photo interpretation, decision about scale of the map to be prepared was taken. The scale depends on the (i) Basic mapping unit, and (ii) the purpose of the map. Here geomorphology is the purpose of the map. Basic mapping unit considered in this case covers an area of 2.5mm^2 on the map and in case of long narrow mapping units, the minimum distance between the two boundaries is 2.5 mm. A base map at scale 1:25,000 was prepared and air-photographs were also used as base maps and the interpretation lines drawn on air-photographs were subsequently transferred on a map by sketchmaster. In interpretating the air-photograph the method advocated by Burring and Vink (1960) was followed and this was discussed in Chapter 2.

In the interpretation of air-photographs for geology and soil science, geomorphic analysis is absolutely necessary. Landform can be directly observed from air-photographs. The most important criterion in geomorphic analysis is the form of the photo image. In geomorphological studies the air photographs are the most useful things affording direct morphological mapping at high level of accuracy.. The wealth of data observed on air-photographs exceed those of topographical map, but they differ in nature. The natural vegetation may give some hints about the prevailing climate. The drainage net work and density also give a clue, but these may be influenced by the lithology. Abrupt change of drainage texture are known with certainty. The tracing of the drainage network may be advantageous by sometimes inverting the photographs to produce pseudoscoping image bringing out the valleys more clearly. This holds good especially for a heavily forested area and for minor streams on flat terrain. In alluvial areas, cut-off meanders and oxbow lakes remain visible on the air-photographs long after their infillings with sediments. Braided channels can be seen long after the river has actually abandoned the course as a whole. In soil mapping a detailed analysis of these channels is rewarding since boundaries are invariably soil boundaries as well. The lateral and down stream shift of free meanders can be measured in this way. On lower terraces often abandoned channels can be recognised.

The following points are important for the study of river terraces.

(1) The thickness of each terrace, deposit and whether nature of the deposit changes downward, gradually or suddenly.
(2) The nature of the deposit, qualitative and quantitative analysis of its pebble or gravel components, measurements of roundness, flatness, elongation indices (morphometric gravel analysis).
(3) In some cases it is advantageous to determine the clay minerals which may change in composition during transport or during sedimentation or even after sedimentation.

In order to interpret the aerial photographs the following procedure have been adopted systematically.

(1) Recognition and identification of the elements.
(2) Analysis of the elements, that is delineation of elements of physical and cultural features of earth surface. This brings out the formation of landform unit (terrain), pattern which refers to the spatial arrangements of objects.
(3) Land classification of the units analysed on air photographs in terms of the physical and cultural characteristics of the earth surface.

For systematic interpretation the following procedure has been adopted.

(1) Study of the index of air-photographs in order to determine the area to be taken up.
(2) Layout of the photographs of scale 1: 30, 000 in their relative position. The layout is then covered with glass plates in order to delineate the main land types, which are described according to general physiography.
(3) A draft classification was carried out in the following way.

    (a) Selection of some sample pairs of photographs from preliminary stereoscopic examination of land types to find the typical characteristics of the land types and the analyse in more detail.
    (b) Selection of other sample pairs from the transition areas between land types and their analysis by stereoscopic examination to delineate the boundaries between the land types and the classify the needed detail in the transition zones.
    (c) Draft classification of all the details delineated on the photographs .

(4) Systematic analysis and classification of other photographs by means of stereoscopic examination and at the sametime classification of the every unit of analysis according to the classification drawn up.

(5) Selection of sample areas and analysis and interpretation of physiographic units in still more detail on enlarged photographs.

After considering all the factors studied on the air-photographs and after field check, a photo-interpretation map (land classification map, based on air -photographs around Rishikesh) has been prepared.

Geomorphic Description

The northern part of the area is a southward extension of a piedmont plain and is covered with thick vegetation. The western portion is a more or less flat area drained by two important rivers, the Song and the Jakhan Rao. The eastern portion is made up, of a series of terraces formed by the Chandna Rao, and the southern portion is occupied by the terraces of the Ganga.

Streams are flowing from north to south and are ultimately debouching in the Ganga. The main streams like the Song have a strongly braided system indicating a sudden decrease of flow velocity after the streams come down from the mountain areas. As the whole drainage system shows a curvilinear pattern towards the west due to the physical barrier of the Siwalik hills of Hardwar Region.

The area has been divided into five geomorphic units based on air-photographs and field work (1) Piedmont plain (2) Chandna Rao terraces (3) Ganga terraces (4) Jakhan Rao — Song interfluve and (5) Song terrace.

The Piedmont Plain

It is bounded on the south by 375 m contour line and is drained by three important streams (meandering) namely the Lalpani Rao, the Bangali Rao and the Gulapani Rao. Each stream possesses a number of tributaries extending northwards by headward erosion forming sub-parallel drainage patterns. The gullies are 6 to 20m deep relative to the piedmont plain. The three main streams have also incised, forming true valley with low terraces bounded by escarpments. These escarpments are not found as soon as the river enters on the terrace land of the Ganga. The three longer streams show deposition of gravels of different sizes in their beds. The shape of the gravels is sub-angular to rounded. Sand is also deposited along the narrow valley floor.

Chandna Rao Terraces

The river Chandna Rao forms a series of terraces, that can be identified with the help of air-photographs and field work. Four terraces have been recognised which gradually step down towards the present valley floor. The terraces are not always bounded by sharp escarpments and do not extend uniformly along the river. They are broken into different parts. The youngest terrace where it crosses the Rishikesh — Dehra Dun Road, is more than 20m above the present valley floor. All the gravelly terraces are covered with a thin veneer of coarse loamy sediments occasionally marked by boulders on flat surfaces.

Ganga Terraces

The river Ganga gives birth to many terraces which are most extensive on the right bank. Most of the terraces are non-cyclic and non-paired due to assymetrical development of the valley profile and

LANDFORM MAP OF RISHIKESH

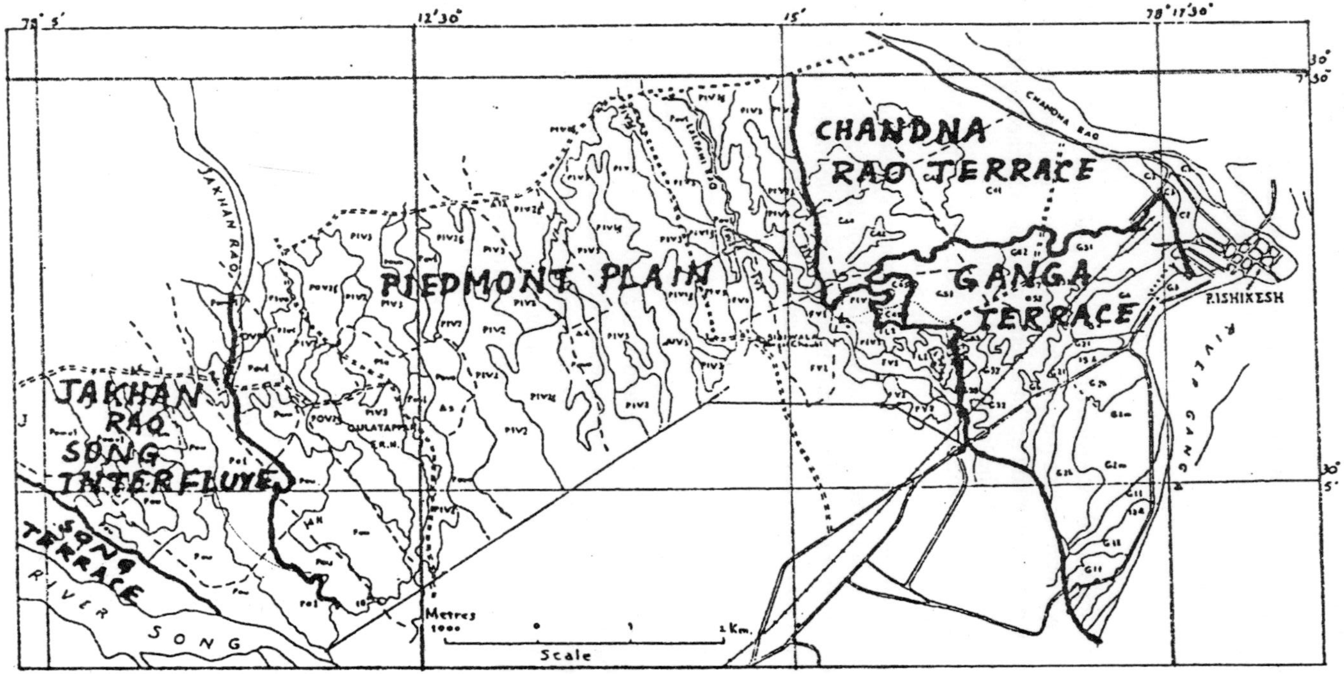

FIG. 5.1 : *Landform Map of Rishkesh*

Source : Paper read in 21st IGU 1968 by Sri R. K. Banerjee and published in 1972 in selected papers Vol-4.

Figure : Classification of landforms of area west of Rishikesh—(GI) Lowest terrace level cultivated, (G11) Light grey and white tone cropped land, (G12) Medium and light tone scattered trees, (G2) Terrace higher than G1, then vegetation and partly cultivated, (G2h) High land, (G2m) Medium land, (G21) Low land, (G3) Terrace higher than G2, (G4) Terrace higher than G3, (G5) Terrace higher than G4, (G51) Light grey and white tone with scattered vegetation and Settlements, (G52) White and dark tone vegetation, (G53) Dark tone with vegetation, (G54) Coarse textured scattered vegetation, (G55) medium textured thin homogeneous vegetation, (G) Terrace bluff, (C) Chandna Rao Terrace landscape forested, (C1) Lowest terrace level, (C2) Terrace higher than C1, (C3) Terrace higher than C2, (C$) Terrace higher than C3, (C41) Dark tone fine textured, (C42) Dark tone medium Texture, (C43) Dark tone fine and coarse texture, (C44) Dark tone coarse texture, (C45) Dark tone coarse texture eroded, (FVI) Fan with light vegetation, (FV2) Fan with moderately uniform vegetation, (FL1) Fan with light tone low lying land, (FL2) Fan with slightly dark tone low lying land, (P1) Higher situation piedmont plain, (PIC) Cultivated, (PIV1) Uniform thin vegetation, (PIV1g) Moderately eroded, (PIV2) Medium vegetation, (PIV2g) Partially eroded, (PIV3) Dense vegetation-uniform texture, (PO) Lower positioned piedmont plain. (PoVo) Unevently distributed vegetation, (PoV1) Light uniform vegetation, (PoV2) Slightly dense vegetation, (PoU) Cultivated plots slightly higher, (PoM) Cultivated plots slightly lower, (PoMV1) Light vegetation, (ST) Song terrace land--partly cultivated, (J) Jakhan-Song Interfluve (partly shown). Profile sites are shown by triangle symbols.

this is the prominent from Lachman Jhula i.e., north of Rishikesh town. The terraces are very prominent and separated by steep bluffs. All the streams of the piedmont region debouch into the Ganga. As the streams come on the higher terraces, they form marshy areas. The water in these marshy and swampy areas comes through the lateral seepage from upper region, particularly form the terraces of the Ganga. The Ganga terraces are formed of fine to medium sand-gravels.

The Jakhan Rao — Song Interfluve

This is located on *Terai* area, which coincides with a cultivated area called Chidrawala and a *Bhabar* tract further west which is partly under a forest plantation called Chandi Block (marked as a part of piedmont) . The Jakhan Rao meets the Song in the extreme south-east of the area. The stream beds possess lots of boulders and pebbles and the interfluve areas are made up of coarse sand to silty materials. In the *Bhabar*, the water is low and the soils are mostly anisotropic sandy and gravelly. In the *Tarai*, the water table is high and the soils are of fine texture. The bed of the Jakhan Rao is very narrow here and no more gravels are not generally found on the surface. This indicates that the river has lost most of its flow due to downward percolation of its water in the gravelly bed further up-stream. In the *Tarai* this water reaches the surface again by underground seepage. This sustained seepage gives rise to the formation of perennial streams which feed the small meandering rivers.

Song Terraces

This area is also located at *Tarai* area as in some places high water table prevails. Numerous small channels of the braiding the Song play an important role in this area regarding deposition of sand, silt and clay. The Song terrace plain is separated from the Jakhan Rao plain by a 3 to 8 m high bluff. This is due to more rapid incision by the Song probably carrying much more water in its lower course.

6

PHOTO-PEDOLOGY OF RISHIKESH AREA

Introduction

Pedological study has been carried out of 40 sq. km. area of Rishikesh Police Station. The study of soil with the help of aerial photographs is known as photo-pedology which is concerned with the study of soils in their natural environment. Soil survey is most typical link between theoretical or fundamental pedology and applied aspects of soil science. The use of soil survey air photographs in soil survey has got positive advantages. First of all it is possible to analyse systematically the whole area to be surveyed with the help of air-photographs. Secondly the vertical air-photographs thus used, the point of view is the same. There are three procedures of soil survey.

(1) Reconnaissance soil survey
(2) Semi-detailed soil survey
(3) Detailed soil survey

Reconnaissance soil survey is nothing but to get a synoptic view the soil of the area. The semi-detailed survey gives the idea of the soils of the area on which planning of detailed soil survey could be made. The detailed soil survey is only applicable to farmlands as number of observations are more numerous than any other soil survey. A comparison of the results of five methods of soil survey are attached herewith. Buring, P. (1960) has recognised five methods of such survey. These are :

(i) A reconnaissance survey, carried out according to the standards of the U.S. Bureau of Land Reclamation and Improvement. One field observation is made on every 100 ha. in a grid system;
(ii) A detailed survey carried out according to the standards of the U.S. Bureau of Reclamation, one field observation is made on every 25 ha. in a grid system;
(iii) A semi-detailed soil survey carried out according to conventional standard for complete field survey, with one observation per 4 ha. using aerial photographs as field and working map;
(iv) A semi-detailed soil survey based on a combination of field observation and aerial photo-interpretation based on 2 observations per 100 ha. is made; and
(v) A semi-detailed soil survey based on method of interpolation described earlier.

So the last three methods belong to semi-detailed soil survey process.

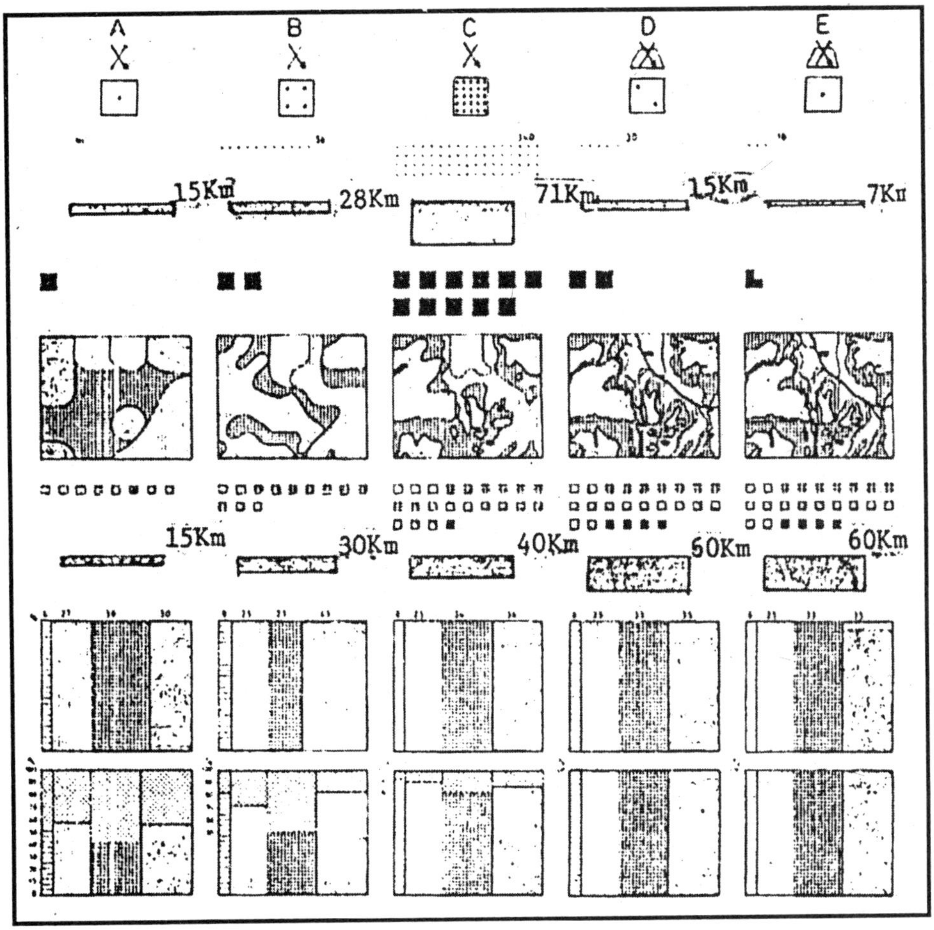

FIG. 6.1 : *Comparison of the Results of Five Methods of Soil Survey*

The description of the procedure undertaken during the preparation of soil map under the scheme of the semi-detailed soil survey is given below :

The stereo-pairs are examined under mirror stereoscope and the terrain is studied in three dimentional aspects and the whole area is scanned on the basis of photo-geography. The basic work of soil study and geographic observation are more or less same because the boundaries of soil are well deducted from specific information of geographical knowledge such as drainage, vegetation, landform etc. But the whole process is a new one which discards the conventional method of all soil survey as the new techniques are meant for preparing maps with the help of aerial photographs. The aerial photographs are taken on the basis of base map in the field. Enlarged rectified photographs are taken into consideration with full annotation. That is the latitudes and longitudes are drawn on the enlarged photographs and rivers and drainage systems are marked, communication systems are drawn clearly on the photographs. At the end, the Enlarged Rectifield Photographs (ERP) are taken as base maps.

Soil map is finally prepared with beautiful description after four hundred augur borings and 10 profile studies. These are discussed at the end of this chapter.

Selection of scale : Before systemetic handling of photo-interpretation for soil mapping, the decision of scale of publishing map has been made, because the scale depends on the following factors —

(1) Basic Mapping Unit,
(2) Purpose of Soil Mapping.

Basic mapping unit considered in the semi-detailed survey covers an area of 2.5 sq. mm. on the map to be published. In case of long narrow mapping unit, the minimum distance between the two boundaries is 2.5 mm. It has been decided that the semi-detailed soil map is usually prepared on the publishing scale of 1:25,000, which is the largest possible scale. It is approximately equivalent to the scale of available photographs. As such, the basic mapping unit is 1.56 ha or 3.74 acres.

A base map is first prepared from aerial photographs supplied by the Survey of India (SOI). The map is prepared by Slotted template i.e. Radial Triangulation method. The results of the photo-interpretation and soil survey from air-photographs are transferred to the base map by the Stereo-pantometer and Sketch master.

After the selection of scale and base map systematic analysis of the physical and cultural features apparent on the photographs are carried out before the actual field work. These physical and cultural features are recognised and analysed according to different photo-elements. In semi-detailed soil study, to interpret the photographs the following steps are adopted.

(1) Recognition and identification of the elements.
(2) Analysis of the elements .
(3) Delineation of physical and cultural features of the earth's surface. This reflects the formation of terrain units or patterns which refer to the spatial arrangement of the object,
(4) Physiographic classification that deals with the classification of units analysed on air-photographs in terms of physical and cultural characteristics of the earth's surface.

For the systematic interpretation the following procedure has been followed :

(1) Studying the aerial photo-index in order to determine the area to be undertaken, because photo-index (P.I.) is most important for classification. Without photo-index or photo-key nothing could be achieved. What the observer wants are detailed in the photo-index (P.I.). This should be carefully preserved and properly oriented for systematic handling of aerial photographs;
(2) Next step is the layout of aerial photographs in the scale of 1:30,000. The layout is then covered with glass plate in order to delineate the main land types, which are described according to general physiography. After preliminary survey, it has been seen that a little larger aerial photographs are needed to describe at least the paleo and quarternery terraces of Recent formation of the Song and Jakhan Rao streams.
(3) A draft classification was carried out as follows : -

    (a) Selection of some sample pairs of aerial photographs from each preliminary landtype for stereoscopic examination to find typical characteristics of land types and to analyse them in more detail.
    (b) Selection of other sample pairs from transition types and their analysis by stereoscopic examination to delineate the boundaries between the landtype and to classify the needed detail in transition areas .
    (c) Draft classification of all the details are delineated on the photographs.

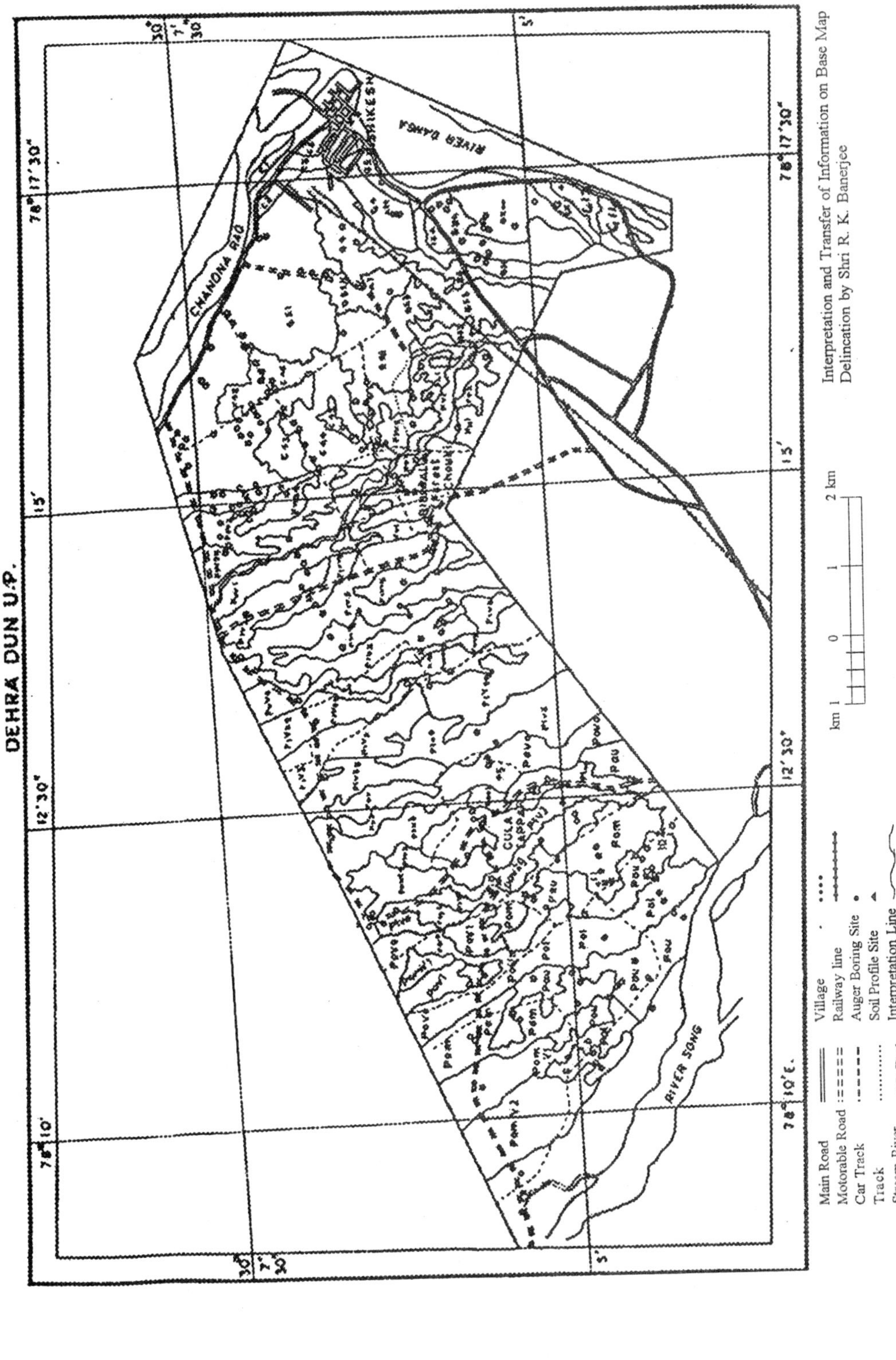

Interpretation and Transfer of Information on Base Map
Delineation by Shri R. K. Banerjee

FIG. 6.2 : *Aerial Photo-interpretation Map of Rishikesh Area, Dehra Dun*

LEGEND OF FIG. 6.2

G-GANGA LANDSCAPE
- G1-Lowest level terrace
  - G11 Light grey and white tone-crop land
  - G12 Medium & light tone-scattered trees
- G2-Terrace higher than GI
  - G2h High land
  - G2m Medium land
  - G2L Low land
- G3 Terrace higher than G2
- G4 Terrace higher than G3
- G5-Terrace higher than G4
  - G51 Light grey & White tone with scattered vegetation and settlement
  - G52 White and dark-tone with vegetation
  - G53 Dark tone with vegetaiton
  - G54 Coarse textured scatterd vegetation
  - G55 Medium textured thin homogeneous vegetation
- G6-Terrace bluff

C-CHANDA RAO LANDSCAPE
- C1 Lowest terrace level
- C2 Terrace higher thanC1
- C3 Terrace higher than C2
- C4 Terrace higher than C3
  - C41 Dark tone fine texture
  - C42 Dark tone medium texture
  - C43 Dark tone fine & Coarse texture
  - C44 Dark tone coarse texture
  - C45 Dark tone coarse texture (eroded)

F-ALLUVIAL FAN
- Fv1 Thin vegetation
- Fv2 Moderately homogeneous vegetation
- FL1 Low and light tone
- FL2 Low land dark tone

P-PIEDMONT PLAIN

P1-Piedmont plain higher situation
- P1c Cultivated and open
- P1v1 Thin homogeneous vegetation P1 V1g-eroded
- P1v2 Moderately dense vegetation P1 V2g-eroded
- P1v3 Dense vegetation fine texture piedmont plain lower situation

Po-Piedmont plain lower situation
- POv0 Scattered and patchy vegetation
- POv1 Thin homogeneous vegetation
- POv2 Moderately dense vegetation
- POh Cultivated higher situation
- POm Cultivated-intermediate situation
  - POMV1 Thin vegetation
  - POMV2 Moderately dense vegetation
- POL Cultivated lower situation

FIG. 6.2 : *Aerial Photo-Interpretation Map of Rishikesh Area, Dehra Dun, U.P.*

(4) Systematic analysis and classification of rest of the photographs by means of stereoscopic examination and at the same time classification of everv unit of analysis according to classification drawn;
(5) Selection of sample areas and analysis and interpretation of hysiographic units in still more detail on enlarged photographs. These enlargements are almost double the scale of the original photographs or about 1:12,500 scale;
(6) Preparation of an interpretation map for soil survey by the transfer of the result of the interpretation to the base map of 1:25,000 scale. The Aerial Photo-interpretation map has been prepared according to the following legend for soil survey.

Legends

Ganga Landscape

G_1 .... Lowest level terrace
- G_{11} — Light gray and white tone crop land
- G_{12} — Medium and light tone scattered trees

G_2 .... Terrace higher than G_1
- G_{2h} — Highland
- G_{2m} — Medium land
- G_{2L} — Lowland

G_3 .... Terrace higher than G_2
G_4 .... Terrace higher than G_3
G_5 .... Terrace higher than G_4
- G_{51} — Light grey white tone with scattered vegetation and settlement
- G_{52} — White and dark tone white vegetation
- G_{53} — Dark tone with vegetation
- G_{54} — Coarse textured scattered vegetation
- G_{55} — Medium textured thin homogeneous vegetation

G_6 .... Terrace bluff

G_c - Chandna Rao Landscape

C_1 .... Lowest terrace level
C_2 .... Terrace higher than C_1
C_3 .... Terrace higher than C_2
C_4 .... Terrace higher than C_3
- C_{41} — Dark tone fine texture
- C_{42} — Dark tone medium texture
- C_{43} — Dark tone fine and coarse texture
- C_{44} — Dark tone coarse texture
- C_{45} — Dark tone coarse texture (eroded)

F - Alluvial fan

FV-1........ Thin vegetation
FV-2........ Moderately homogenous vegetation

FL-1........ Lowland light tone
FL-2........ Lowland dark tone

P - Piedmont plain

P 1 — Piedmont plain higher situation
- P 1C — Cultivated and open
- P 1 Vig — Eroded
- P V2 — Moderately dense vegetation
- P V2g — Eroded
- P 1 V3 — Dense vegetation fine textured

PO - Piedmont plain lower situation

PoVO - Scattered and patchy vegetation
PoV 1 - Thin homogenous vegetation
PoV 2 - Moderately dense vegetation
Pou - Cultivated high land
Pom - Cultivated intermediate situation
- PoMV¹ - Thin vegetation
- PoMV² - Moderately dense vegetation

PoL - Cultivated

Soil Survey Procedure

The procedure of the interpretation with full check has been followed during the field work for semi-detailed soil survey. In this procedure the systematic interpretation of air-photographs is conducted before the field survey. The sample areas are selected in order to carry out the detailed soil survey for classification of soils into families and phases. Detailed soil survey of the sample areas through field check has been made. The survey was carried out by aerial photographs of 1:25,000 scale. Sometimes it is not possible to make clear out boundaries between families and phases and the soil map produces association of families and phases. The selection of sample areas plays an important role in the "Interpretation with full check" procedure. The conditions for selection of sample areas are :

(1) They are typical for a large area
(2) They include all soils occurring within the study area
(3) They are easy to access

The principle is to know the details of the soils in the sample areas and to extend this knowledge and experience to surrounding areas with the help of information provided by the stereoscopic photo-image. After completion of the photo-interpretation map, field work is carried out first in the sampling areas after which a first draft classification of the soils could be made. It was followed by a first field survey (check survey) in the whole area. Subsequently a draft legend of the soil map and a partial draft of soil map are made. In the second field survey the remainder of the survey is to be carried out and the draft legend is made final alongwith a first general draft map and additional notes on dubious points. In the third field survey (revision survey), spot visits on dubious points have been made for the final draft soil map and revised legend.

The results of investigation are interpolated on air-photographs for subsequent transfer on a base map of 1:25,000 scale. It is impossible to give quantitative indications on the time needed for various steps of soil survey. This differs according to the survey procedures followed by selecting the scale of the soil map and to the physiography of the area and its correlation with the soil units. The following steps should be taken as indicative of the gradual grouping of air photo-interpretation field observation and deductive to form a soil map of good quality. This diagram also demonstrate that legend should be made during the field work. Any attempt to make a legend after field work has been completed, give rise to unexpanded problems and to a soil map of poor quality.

Table 6.1 : Sequence of Laboratory — Field Experiences in Remote Sensing

Place	Phase	Result
Laboratory	1. Air-photo-interpretation	1. Photo-interpretation map with (physiographic classification)
Field	2. First Field Reconnaissance	2. First draft legend of the soil map.
In Field Camp	3. Mapping of Sample areas or catenas	3. Second draft of the soil map plus first partial draft of the soil map
Field Camp	4. First Field Survey	4. Checked field legend made final plus second partial draft of soil map
Second Field	5. Second Field Survey	5. Remainder of the survey carried out with final field legend plus first general draft of soil map plus additional notes on dubious points
Third Field	6. Third Field Survey (Reservation survey)	6. Spot visits on dubious points (where possible lead to final draft soil map plus revised field legend)
Laboratory	7. Map editing plus draft report	7. Final polishing of map plus careful editing of final survey legend plus the survey report which is as essential as the soil map itself.

Source : Planning of Soil Surveys in land development, Vink 1963.

Soil Genesis and Classification

The soils of the area have been classified on the basis of the Seventh Approximation of United States Department of Agriculture (USDA). Subsequently, it is known as Taxonomic Classification, which pays emphasis on natural relations that exists between the soils, reflected in classification groupings.

The Seventh Approximation Classification System is based on pedo-genesis and follows two general principles which are as follows :

(1) to study internal properties of the soil profiles themselves. Outside influence affecting the properties are not used as far as possible (climate).
(2) use of well defined morphological and measurable chemical, physical, mineralogical and biological properties of the profile.

The two principles lead to the adoption of the following criteria used in the classification system:

(a) The stage of development of profile.
(b) The type of alteration.
(c) Type of organic matter and its influence on soil formation.
(d) Hydro-morphology of the profile.

The four criteria mentioned above are generally followed. The new concept of the 7th approximation classification system is to identify the soil horizons.

(1) Epi-pedons or diagnostic surface horizon.
(2) Diagnostic sub-surface horizon.

These horizons are genetic soil horizons with well defined properties. Before deciphering soil genesis and classification, the smallest unit of the soil need to be known on a scientific basis. The smallest unit of a soil is known as 'PEDON'. Some soil properties can be determined from the surface to determine the nature of a soil as well as its horizons or layers. This study requires pits or some means of extracing soil samples from the surface to the base of the soil. The visible or tractile properties of samples can be studied in the field. Soil moisture or temperature regions are studied by observation of changes over time at representative points. Other properties of a soil must be learned by studies of samples in a laboratory.

A PEDON is the smallest unit or area for which we should describe and study the sample of the soil to represent the nature and arrangement of its horizon and variability of other properties that are preserved in samples. A Pedon is comparable in some ways to the unit cell of a crystal. It has three dimensions. Its lower limit is somewhat vague and lies between the soil and the hard rock. Its lateral dimension is large enough to represent the nature of any horizon and variability that may be present. A horizon may vary in thickness or in composition or it may be discontinuous. The area of a Pedon ranges from l to $10m^2$, depending on its variability in the soil. The uppermost layer of the soil is known as epi-pedon. There are may kinds of epi-pedons and surficial properties. The following epi-pedons areas are recognised in the study area.

1. *Mollic epi-pedon :* Dark coloured, thick, humiferous rich in bivalent cations, base saturation is above 50 per cent, gradual transition to underlying horizon, colour darker than C or underlying horizon, it contains 1 per cent organic matter than normal. The thickness of mollic epi-pedon will be the thickness of fine earth till the coarser fragments are removed.
2. *Ochric epi-pedon :* This is poorer in humus with lighter colours than mollic epi-pedon, rubbed colour value of 5.5 or higher on dry or 3.5 or higher on moist. The only diagnostic subsurface horizon found in the study area is *Cambic horizon* — Texture finer than loamy fine sand; formed under two different conditions (1) well drained, aerated initial development of a horizon due to combination of complete leaching of lime, liberation of free acid from oxidation (brownification) and clay, formation of soil structure in aerated condition. (2) Indrained reduced condition — In anaerobic condition on the horizon just below the epi-pedon is dominated by low chroma or by blue and green hues (60% of the mass) in the matrix peds are absent. The exact criteria are (a) chromas of 2 or less in values of 4 or more of hues are red or redder than 10 YR or bluer. There are ten orders in the 7th Approximation of soil classification, subsequently known as Taxonomic classification of USDA.

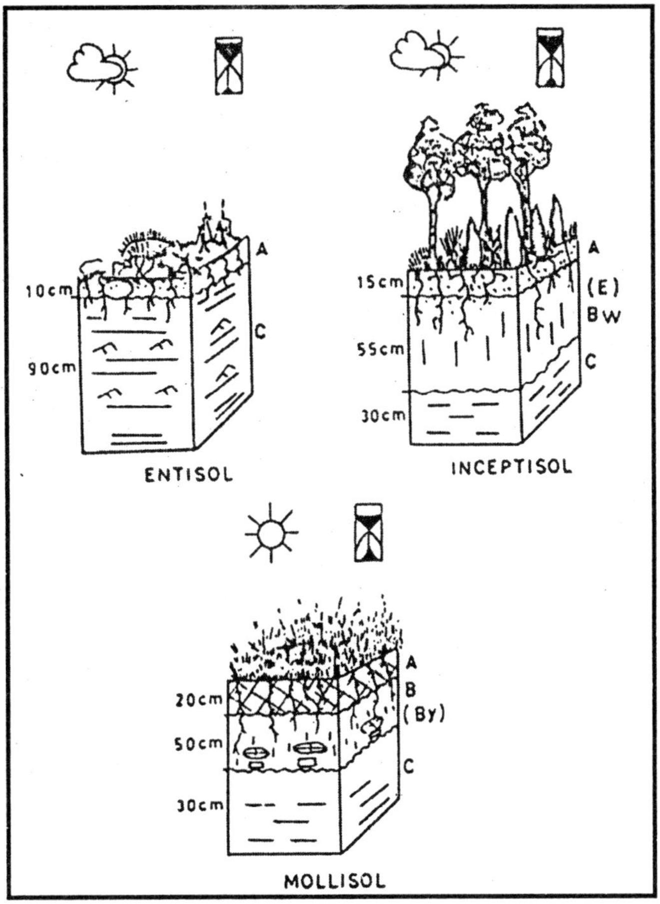

FIG. 6.3 : *Vegetation and Soil Profiles of orders of Entisol, Inceptisol and Mollisol*

1. Entisols
2. Vertisols
3. Inceptisols
4. Mollisols
5. Spodosols
6. Oxisols
7. Alfisols
8. Ultisols
9. Histosols
10. Aridisols

Out of these 10 orders, only three orders are found in the study area. These are — 1. Entisols (which means recent soil and soil profile are immaturely developed) 2. Inceptisols (Matured profile) and 3. Mollisols (soft blackish surface and matured profile) .

Suborders are also defined with the help of diagnostic horizon, but mainly by the internal micro-climate of the soils and hydromorphy. The period of total wetness is important. Therefore, many orders, suborders exists for —

1. Permanent wet soil
2. Moist soil that is not saturated for a long period.
3. Dry soils

An example from this area is that the division of inceptisols may clearly be made on the basis of moist and dry condition as *Aquetps* and *Ochrepts.* Entisols may form a suborder by presence of sands permanently noted as Psamment.

The great groups are defined by the presence or absence of diagnostic horizon and the arrangement of the horizon. As it is known usually that the name of a great group consists of one or two formative elements suggesting something of the diagnostic properties. Names of great group, therefore, have three or four syllabus and end with the name of a suborder. For example, Psamments, are extreme condition of materials; as such, the name of great group Psammaquents. Psamments is the suborder and Psammaquents belong to the great groups. The chart shows the formative elements of soil order from which suborders, great groups are derived.

Table 6.2 : Formative Elements of Soil Order

Name of order	Formative element in the name of order	Derivation of formation elements	Pronounciation of formative element
1. Alfisols	Alf	Meaningless syllable	Pedelfar (acid soil)
2. Aridisols	Idl	L. Aridus, dry	Arid
3. Entisols	Ent.	Meaningless syllable	Recent
4. Histosols	Ist.	Gr. Histos, tissue	Histology
5. Inceptisols	Ept.	L. Inceptus, beginning	Inception
6. Mollisols	O11	L.Mollis, soft	Mollify
7. Oxisols	Ox	F.Oxide, Oxide	Oxide
8. Spodosols	Od	Gr. Spodos, Wood as L	Odd
9. Ultisols	Ult.	L. Ultimus, last	Ultimate
10. Vertisols	Ert	L. Verto, turn	Invert

Source : Soil Taxonomy, USDA, 1975.

Further sub-groups are sub-division of great group and can be defined as —

1. The Central concept of the great group
2. Intergrade to other orders, suborders or great group as reflected by the properties of the other class either as :

   (a) A merging of properties throughout the soils
   (b) Additional horizon (say buried horizon)
   (c) Intermittent horizon.

The nomenclature :

Alf (Alfisols); Id (Aridisols); Ent (Entisols); Hist (Histosols); Ept (Inceptisols); Oll (Mollisols); Ox (Oxisols); Od (Spodosols); Ult (Ultisols); Ert (Vertisols) and by a prefix that indicates the characteristics of the suborder being very often micro-climate.

Names of the suborders according to diagnostic horizons, micro-climate characteristics and extreme kinds of parent material accuring in the study area given in the following table :

Table 6.3 : Diagnostic horizons, Micro-climate Characteristics and Parent Materials

Order	Characteristic of suborder	Suborder name
According to Diagnostic Horizon		
Ent (Entisols)	Fluv (indication of alluviation on slope less than 25%	Fluvent
Ept. (Inceptisols)	Ochr (Ochros — light coloured diagnostic epipedon)	Ochrept
According to Parent Material		
Ent (Entisols)	Psamm (sandy)	Psamment
According to extreme kind of parent material ,		
O11 (Mollisols)	Ust (dry soil climate)	Ustoll
Ept (Inceptisols)	Aqu (hydromorphic condition)	Aquept

Source : Soil Taxonomy USDA 1975

Table 6.4 : Names of a great Soil Group, their suborder : A Prefix refering to diagnostic horizon

Order	Suborder name	Great Group	Remarks
Ent.	Fluvent	Usti-flu-vent (dry soil climate) Xero-psamment (very dry soil climate) Aqu-psamment (horizon mottled or completely reduced under wet condition).	(Entisol has no diagnostic horizon. Hence fluvent and psamment are grouped into great group by micro-climate condition).
Ept.	Ochrept	Ust-ochrept	(Cambic horizon under dry condition).
	Aquept	Hapl-aquept	Hapt-simple diagnostic mottled cambric horizon.
	Hst-Oll	Hapl-ust-Oll	Simple cambric horizon under dry aerated condition, underlying mollic epipedon.

Source : Soil Taxonomy USDA, 1975.

Table 6.5 : Names of Soil Sub-group with an Adjective Indicating Direction of Pedogenesis

Order	Suborder name	Great Group	Remarks
Ent	Fluv-ent	Usti-fluv-ent	Aquic usti-fluvent (Pedogenesis directed by hydromorphic). Typic usti-fluv-ent Typic xeropsamment. (Central concept of the group).
Ept.	Ochrept	Ust-Ochr-ept.	Typic Ust-Ochrept (Central concept of the group).
	Aqu-ept	Hapl-aqu-ept.	Typc hapl-aqu-ept (Central concept of haplaquept). Aeric hapl apl-aqu-ept (Pedo-genesis directed towards partial reduction and oxidation condition).
O11	Ust	Hapl-ust-oll	Mollic hapl-aquept (inter-grade to other order indicating direction towards development of mollisols) Udic hapl-ust-oll (deviation from Central concept with the subgroup. Aquic udic-hapl-ust-oll (intergrade tendency — properties not indicative of any other great group).

Source : Soil Taxonomy USDA 1975

Soil family is the sub-division of the sub-group on the basis of properties important to the growth of plant. The differentiating properties are chosen so that the family is uniform with respect to fertility. This reflects the Solum (A+B) below plough depth. Criteria of classification of families used in the study area are —

1. Particle size distribution in horizon of major biologic activity below plough depth;
2. Mineralogy of the same horizon that are considered in naming particle size classes.
3. Symptom of the soil that is pH values
4. Temperature regime
5. Thickness of soil penetrable by roots.

The above mentioned criteria are properties that are relatively permanent. Arbitrary control section for classifying texture classes is between 25 cm below the surface and 100 cm or a lithic or a paralithic contact, whichever is shallower.

Criteria of Classification of Family Grouping

A. *Texture classes for family grouping*

1. Sandy skeletal : More than 50 per cent volume, coarser than 2mm with enough fines to fill interstices larger than 1mm fraction, less than 2 mm is defined for texture class 5.
2. Sands and loamy sands coarser than loamy very fine sand.
3. Loamy

   (a) Coarse loamy soil with less than 18 per cent clays.
   — Coarse sandy loam

— Sandy clay loam
— Fine sandy loam
— Loam

With less than 18 per cent clay and more than 15 per cent coarser than fine sand (including coarse fragments up to 2 mm)

Very fine sandy loam
Loamy very fine sand
Silt loam

(b) Fine loam with more than 18 per cent clay
— Coarse sandy loam
— Sandy loam
— Fine sandy loam

With more than 18 per cent clay but less than 15 per cent clay and more than 15 per cent coarser than very final sand including coarser fragment up to

— Very fine sandy loam
— Silt loam
— Sandy clay loam
— Silty clay loam

(c) Coarse silty loam with less than 18 per cent clay and less than 15 per cent coarser than very fine (including coarse fragments up to 2 mm.)

— Loamy very fine sand
— Very fine sandy loam
— Silt loam
— Silt

(d) Fine silty loam with more than 35 per cent clay and less than 15 per cent coarser than very fine sand including coarse fragments of 2mm.

Very fine sandy loam, silt loam,
silt clay, coarser loam, clay loam

B. *Soil mineralogy classes for family grouping*

(a) *Micaceous* — In soils with texture coarser than clay more than 40 per cent micas by weight of clay size and coarser, but excluding fragments coarser than 7.5 per cent for clayey soils less than 35 per cent of volume of fragments coarser than 2 per cent the amount of clay, fine earth fraction in 35 per cent or more by weight.

(b) *Calcareous* — Used to indicate the continuous presence of free carbonates in all parts of fine earth fraction between depths of 25 cm - 50 cm in Aquepts other than Fragiaquepts in Aquoll and in Entisols other than Psamments.

C. *Reaction classes*

By pH values the symptom of soils are determined whether a soil is acid or basic.

(a) Acid pH less than 5.5 per cent throughout the *Control Section :* generally the depth of the control section is considered up to 100 cm depth of the soil.
(b) Non acid pH 5.5 or more in at least some parts of control section.

Reaction classes are to be used in the following taxonomic classification.
Entisols
Haplaquepts
Thus, we have the following combination reflecting the pH in the control section.
Entisols
Haplaquepts
The connotation of the term *Control Section* is given below.

Names of particle size classes or their substitutes as defined to fragipan, duripan or a petro-calcic horizon but applied to specific horizon or to the materials between given depth that are defined in terms of either the distance below the surface of the mineral soil or the upper boundary of the specific horizon. The vertical section so defined is called *Control Section.* Definition of control section for determination of the particle size classes are arranged as a key. The vertical section for studying the soil depends upon the underlying surfaces and to some extent the geographical condition such as micro-relief variations. Generally it varies from 10 cm. to 1 metre in depth.

During the field study, the following combination reflecting the *Control Section* and presence or absence of carbonates in all parts of fine earth between 25 and 50 cm below the surface have been observed.

Psamments	Acid and non-acid classes
other Entisols and aquepts except Fragiaquept and unbrepts.	Acid and non-acid and calcareous classes (note that both names are not needed for calcareous families).

D. *Soil depth classes*

(1) *Shallow :* Less than 50 cm to base of soil used in all great group of Entisols, Inceptisols, Aridisols, and Ultisols except cryic group (ocrya aquepts, only umbrepts and so on).
Note : That the ochric soils and lithic subgroups are also shallow, but adjective shallow in the family name is redundant.

(2) *Deep :* Uptil now not defined. The normal range of thickness of soil varies from one great group to another and will need to be established. ‘ Deep’ is the term and fixation is being done as per field work about the fixation of the scale.

(3) *Very deep :* More than 1 metre thickness of soil. During photo-pedological study of the

area, the orders of the soils are not according to the 7th Approximation of recent soil classification. Latest soil classification is known as Soil Taxonomy (a basic system of soil classification for making and interpreting soil surveys, 1975) . The aerial photographs were taken as the base map and picked the points where detailed study was made in the sample area on the air-photographs. The sample area consists of 10 per cent of the study, where detailed study was carried out. The difference of detail and semi-detailed survey are illustrated here to understand the situation. After careful and critical study on air-photographs, Satellite Imagery and field checkings, the following soils are found :

1. Entisols
2. Inceptisols
3. Mollisols

Entisols

Entisols occur in the younger terraces of the Ganga, the Chandna Rao and the Song rivers and other minor tributaries of the Ganga. They are characteristed by mineral soils, which have no diagnostic horizons other than an Ochric or an anthropic epi-pedon, an albic or an agric horizon. Entisols are either without natural genetic horizons or with only the beginning of the horizons. The horizons are so weakly developed that they fail to meet the requirements of any of the diagnostic horizon except the albic horizon. These are produced through cultivation by man. Entisols of Hardwar Rishikesh area are found on recent calcareous alluvium. They normally show textural layering and lamination which have not been completely destroyed by biological action. The structures of the layers are generally single grained to massive or only weak subangular blocky, indicating very poor profile development.

The Entisols of the area have been sub-divided into the following suborders as the development of this order is limited to two suborders due to climatic and biological factors.

1. Psamments
2. Fluvents

1. *Psamments :* Entisols occurring

(1) below the A_p horizon or 25 cm, which may be even deeper, texture of loamy fine sand, coarser in all parts either to a depth of 1 metre or to a lithic or paralithic contact, which ever is shallower.

(2) Lack fragments of diagnostic horizon, more or less without discernable order in the soil below A_p horizon.

Psamments have further been subdivided into great group as follows :

1. Xeropsamment
2. Aquipsamment

(i) *Xeropsamment :* Psamment that (1) have a mean soil temperature of 8.3°C or more, but less than 22°C and mean summer and mean winter soil temperature, at 50 cm that differ by 5°C.

(ii) are usually moist but are dry for more than 60 consequetive days in more than 7 out of 10 years in all subhorizons between 25 cm and 1 metre or a lithic or a paralithic contact, which ever is shallower.

(iii) have in the sand fraction less than 95 per cent quartz, zircon, tourmaline, rutile or other normally insoluable mineral that do not weather to liberate iron or aluminium.

(iv) are not saturated with water at any season or lack the characteristics, associated with wetness defined for aquipsamment.

We have found only one sub-group of Xeropsamments, viz., the Typic Xeropsamment.
Typic Xeropsamment : are xeropsamments that (i) lack lammellae within 2 metres of the soil surface, that meet all requirments for an argillic horizon except thickness. (ii) have no mottles with chromas of 2 or less to a depth 1 metre; (iii) lack within 1 m of the surface any horizon that is more than 15 cm thick, that contains either at least 20 per cent durinodes in nonbrittle and has firm consistency.

Typic Xeropsamments have besn subdivided in family as follows :

1. Non-Acid

    Non-micaceons
    (a) Shallow (on sandy skeletal)
    (b) Deep (on sandy skeletal)
    (c) Very deep.

    Micaceous
    (a) Shallow (on sandy skeletal)
    (b) deep (on sandy skeletal)
    (c) very deep. .

2. Calcareons (non-micaceous)
    (a) Shallow on sandy skeletal typical profile description of a very deep non-micaceous typic xeropsamment is given.

Table 6.6 : Profile character of the soils at Chidrawala village Pit No. 11

landform - Jaklan - Son Interfluve
Relief - almost flat and level
Landuse - partly cultivated

Depth (in cm)	Characteristics
0-7	Dark grayish brown (2.5Y 4/2) light brownish gray 2.5Y 6/2 on dry, loamy sand, single grained, loose, very friable warm casts and termite excrements, non-acid, abrupts smooth boundary.
7-22	Dark grayish brown (2.5Y 4/2) loamy sand (heavy); massive to very weak.
22 -24	Grayish brown to brown (2.5Y 4/2) to (l0YR 5/3) anisotropic loamy sand (heavy and light), massive to single grain, very friable, on friable on moist, non-sticky on wet; few termite excrements, few large and common pores, non-acid, abrupt and irregular boundary.
40-64/80	Brown to light grayish brown (2.5Y to 10 YR 5/3) massive, very friable on moist non-study on wet, few work casts and termite excrements, few coarse and common medium pores, non-acid, abrupt and irregular boundary.
64/80-100	Grayish brown (2.5Y 4/2) loamy sand (heavy) massive, very friable on moist and non-sticky on wet, few worm casts and excrements, few coarse and common medium pores; non-acid; gradual and smooth boundary.
100 -118	Grayish brown (2. 5Y 4/2) loamy sand ; common reduction (2. 5Y b /o) and oxidation (l0YR 5/4) mottles ; massive, very friable on moist and non-sticky wet; few worm casts and excrements; few medium pores; non-acid.

Note : For genetic classification control section of 1 m is only considered

(*Source* : Prepared from boring results of field survey)

(ii) *Aquipsamments*

(1) Psamments which have a mean annual soil temperature of more than 8.3°C and a mean summer soil temperature at 50 cm or at lithic or aparalithic contact, whichever is shallower of either more than 15.5°C of cultivated or without an 0 horizon or more 8.3°C if within 0 horizon. It is found during field study that recent cultivated area does not possess any top horizon and does not develop any agricultural horizon as in the western countries. Land put under prolonged agricultural use experiences mulching, the agriculture horizon fully developed may be considered A_{00} horizon.

(2) are saturated with water at some season of the year or are artificially drained and that have depths of left that 50 cm moist colours associated with wetness namely.

(a) If hues are as red or redder than l0YR and there is mottling chromas 2 or less, or there is mottling and values are less than 4 and chromas are less than 1 or values are 4 or more, chromas are 1 or less.

(b) If hues are between l0YR/10Y and there is distinct or prominent mottling, chromas are 3 or less; if there is no mottling, chromas are 1 or less.

(c) Hues are bluer than 10Y.

(d) Any colour which is due to uncoated grains of sand Aquipsamment have been further subdivided into a subgroup which is typically called Typic Aquipsamments.

Typic Aquipsamments are aquipsamments

(a) which have chromas of 2 or less in 80 per cent of the mass below any A horizon and down to depth of 75 cm or more; the colour is due to uncoated sand grains; ground water table is within 50 cm of the surface for more than of the time.

(b) which have no lithic contact within 50 cm of surface.

(c) which have less than 5 per cent by volume non-indurated plinthite in all horizons to a depth of 1.25 m.

(d) which have no surface horizon that meets all the requirements of mollic (soft) epitedon except thickness.

Typic Xeropsamments have three families as —

(1) Shallow (on sandy skeletal) non-acid.
(2) Deep (on sandy skeletal) non-acid.
(3) Very deep, non-acid.

Table 6.7 : Profile Character of Typic Aquipsemment at Chidrawala (angur boring results)

Landform : Jakhan - Song Interfluve
Relief : almost flat & level.
Landuse : Agriculture.

Depth (in cm)	Character
0 - 12.5	Dark gray (l0YR 4/1) loamy sand or sandy loam with rusty specs of dark yellow brown (l0YR 4/4) friable, non-sticky, weak effervescence with hydrochloric acid.
12.5 - 25	Dark gray (10YR 4/1, 80%) mottled with dark yellowish brown (l0YR 4/4) loamy sand, friable, non-sticky weak effervescence
25 - 50	Gray (2.5YR 5/0, with faint mottlings of light olive brown to yellowish brown (2.5Y 4/4 to YR 5/8 on exposure, sand, loose, non-acid.
50 - 65	Loose gravels, pebbles and cobbles, rounded.

Source : Prepared from boring results of field summary.

Fluvents are the suborders of entisols. They have the following characters :

(1) textures of loamy very fine sand or finer in some horizons below the A horizon or 25 cm whichever is deeper, but above a depth of 1 metre or lithic or paralithic contact whichever is shallower.
(2) lack fragments of diagnostic horizon that occur more or less without discernible order in the soil below Ap horizon.
(3) They have slopes of less than 25 per cent.
(4) They have an organic matter content that decreases regularly with depth or that remains above 0.35 metre. Thin strata of sand or loamy sand may have organic matter of the finer sediments at 1.25 m or below have 0.35 per cent (0.2 per cent carbon) or more. Fluvents of the area studied have one great group, known here as Ustifluvents. It has been established that only this category contains subgroups.

(1) Ustifluvents are fluvents that —

(1) have soil temperature similar aquipsament.
(2) are usually moist.
(3) are dry for more than 90 cumulative days in the moist years in some sub-horizon between 25cm and 1 metre or a lithic or paralithic contact, whichever is shallower but are not continuously dry in all sub-horizons between these depths for as long as 60 consecutive days in more than 7 out of 10 years, unless either the mean annual soil temperature is 22° C or more or the mean summer and mean winter soil temperature at 50 cm depth differ by less than 3°C or both. In this area ustifluvent have two sub-groups :-

(1) Typic Ustifluvent.
(2) Aquic Ustifluvents.

Typic Ustifluvents are ustifluvents which —

(a) lie within 50 cm of the surface, mottles with chromas of 2 or less at depths within 1.5 metre of the surface, lack horizons that are saturated with water at some season or that are artificially drained and that have chromas of less than 1 or hues bluer than 10Y.
(b) lack the following characteristic (1) cracks at some season in most years that are 1 cm or more wide at a depth of 50 cm and that are at least 30 cm long in some parts.
(c) lack a lithic contact within 50 cm thick, of the surface.
(d) lack within 1 metre of the surface any horizon that is more than 15 cm thick, containing either at least 20 per cent durinodes in a non-brittle matrix or its brittle has firm consistence when moist.

Typic Ustifluvents are non-acid and have been subdivided into :

(1) Coarse loamy
(2) Coarse silty
(3) Coarse silty
(4) Sandy

Coarse loamy typic ustifluvents, non-acid have again been sub-divided according to the depth of solum as :

(1) coarse loamy
   (a) deep (on sandy skeletal)
   (b) very deep.

Similarly coarse silty ustifluvent non-acid have also been subdivided according to the depth of the solum as :

(2) coarse silty
   (a) deep (on sandy skeletal)
   (b) very deep.

(3) Coarse silty and loamy anisotropic non-acid type; ustifluvents are very deep.
(4) Sandy typic ustifluvents non-acid are deep on anisotropic layer of sandy loam and silty loam.

Table 6. 8 : Profile Character of Coarse Loamy Typic Ustifluvents of Chidrawala

Land form : Jakhan — Song Interfluve Relief
Relief : almost flat level plain
Land use : agriculture

Depth (in cm)	Characteristics
0 - 12.5	dark grayish brown (10YR 4/2) sandy loam friable on moist and non sticky on wet, mica flakes common, non-acid.
12.5 - 37.5	dark grayish brown to brown (10 YR 4/2/2.5); sandy loam, friable on moist, non-sticky on wet, mica flakes common; non-acid.
37.5 - 75 cm	dark grayish brown (10YR 4/2/2.5); sandy loam friable on moist, slightly sticky on wet, mica flakes few, non-acid.
75 - 125	dark yellowish brown (10YR 4/3) sandy loam, friable on moist, slightly sticky on wet, mica flakes few non-acid.

Source : Prepared from boring results of field survey.

2. Aquic Ustifluvents are ustifluvents like the typic except the characteristics as mentioned for (a) The sub-group of aquic ustifluvent, non-acid have been subdivided into following families :

(1) Coarse loamy
(2) Coarse silty
(3) Sandy

Coarse silty aquick ustifluvents have also been further sub-grouped into —
(a) deep on sandy skeletal
(b) very deep

Coarse silty aquick ustifluvents have also been further subgrouped into -
(a) deep on sandy skeletal
(b) very deep

Sandy Aquick Ustifluvents are orily deep on anisotropic of sandy loam and silty loam.

Table 6.9 : Profile Character of Aquic Ustifluvent at Chandi Block, forest village

Land form : The Jakhan — Song Interfluve
Relief : Almost flat level plain
Land use : Forest plantation

Depth (in cm)	Characteristics
0 - 10	Grayish brown (2.5 Y 5/2) silty loam, massive with strongly disturbed lamination, slightly firm on moist, moderately sticky on wet, few excrements, very few medium and coarse pores, rusty specks, non-acid abrupt and smooth boundary.
10 - 45/50	Grayish brown to light olive brown (2.5 Y 5/2 to 5/3) silt loam and fine sandy loam in alternating thin layers; faint lamination for heavier material and clear lamination in lighter materials, friable on moist, slightly sticky on wet, common excrements; few medium and coarse pores, many distinct form of mottling particularly in the lamination banding and cross, non-acid, abrupt smooth boundary.
45/50 - 75/80	Dark grayish brown (2.5 Y 4/2) silty loam massive but locally disturbed lamination, friable, moist and sticky on wet, common excrements; common medium and few coarse pores, common faint iron and reduction mottled; non-acid, abrupt and smooth boundary.
75/80 - 96	Pale brown to brown (10 YR - 5.5/3) very fine sandy loam, clear lamination, friable on moist and slightly sticky on wet, few medium and coarse pores, common faint iron mottlings, non-acid abrupt and smooth boundary.
96 - 101	Dark grayish brown to olive brown (2.5 Y 4/3), sandy loam, massive strongly disturbed lamination, very friable on moist and non-sticky on wet, common excrements, common medium and coarse pores, non-acid abrupt and smooth boundary.
101 - 120	Grayish brown to light olive brown (2.5 Y 5/2 and 5/3), silty loam strongly distributed lamination, friable, moist, sticky or wet, many excrements, many and few coarse pores, many faint reduction and iron mottling, non-acid.

Source : Prepared from boring results of field survey.

Inceptisols

The Inseptisols are the soils of the humid regions which have altered horizons. They have lost bases of iron and aluminium but retain some weathered minerals. They do not have an illuvial horizon enriched either with silicate clay that contains aluminium or with an amorphous mixture of aluminium and organic carbon. Horizons of accumulation of translocated silica, iron or bases are permitted. The Inceptisols may have many kinds of diagnostic horizons, but shallow plinthite and natric, argillic, spodic, oxic, gypsic, petrogypsic and salic horizons are excluded. Only very few inceptisols are permitted to have a mollic epipedon.

Inceptisols occur more extensively in the study area particularly in the forested, interfluve, doabs and in the piedmont areas. This type of soil is more mature than the Entisols. The real conventional equivalent of Inceptisols in the Revised classification of Baldwin, Kellogg and Thorp (1960) are ando, Sol Brun Acide some Brown forest soil, low Humic grey soils. The soils occur specially in the *Terai* which occupies in between piedmont plain and oldest river terraces of either the Ganga or the Song river tributaries of the Ganga.

Inceptisols are characterised here by mineral soils which are usually moist between 25 cm and 1 meter or a lithic or a paralithic contact, or shallower than 1 meter. The may have spodic, argillic or oxic horizons, unless it is buried horizon. The calcic horizon does not immediately underlie a calcareous epipedon and which have one or more of the following :

1. (a) Cambic horizon
   (b) an Umbric, histic or plaggen epi-pedon
   (c) a fragipan that has no caly skins as thick as 1 mm
   (d) a duripan

2. A mollic epi-pedon.
   There are one or more of the following combination of properties.

   (a) The base saturation is less than 50 per cent (by NH_4OH) in some part ofthe underlying of the cabmic horizon.
   (b) The mean summer and winter soil temperature differ by less than 5°C when measured at a depth 50 cm or at a lithic or paralithic contact whichever is shallower.
      (1) moist and dry values no darker than those of underlying horizon.

3. Saturation with water at some period of the year when not frozen or artificially drained at depths of less than 50 cm or immediately underlying an umbric epipedon, have a cambic horizon with moist colours that meet one of the following requirements :

   (a) If there is mottling, chromas are 2 or less
   (b) If there is no mottling and values are less than 4, chromas are less than 1, if values are 4 or more, chromas are 1 or less.

Sub-order : There are two sub-orders of Inceptisols, which are —

(1) Aquepts (2) Ochrepts

The *aquepts* are inceptisols which are either saturated with water at some period of the year or are artificially drained; have a depth of less than 50 cm, characteristics associated with wetness namely—

1. Moist colours on ped-faces, if peds are present or dominant colours in the matrix as follows:

   (a) If there is mottling, chromas are 2 or less.
   (b) If there is no mottling chromas are 1 or less ; if hues are redder than 10 YR because of red parent materials that remain red after citrate-dithionite extraction, the requirement for low chroma is waived.
   (c) hues are bluer than 10 Y .

The Aquepts have been further subdivided into great groups as Haplaquepts.
The Haplaquepts are aquepts — which have

(1) a mean annual soil temperature of more than 8. 9°C ; mean summer and winter soil temperature at 50 cm that differ by 5°C or more.
(2) an ochric epipedon
(3) N value of 0. 5 or less in between 20 and 50 cm

(4) lack a fragipan
(5) no plinthite forming a continuous phase or constitutes more than half of the matrix within 1.25 m of the surface.

Haplaquepts have three sub-groups in this region such as —

(1) Typic Haplaquepts
(2) Aeric Haplaquepts
(3) Mollic Haplaquepts

Typic Haplaquepts

The central concept and sub-groups of Haplaquepts are fixed on dominantly grey soils in thick mineral regolith without deep wide cracks at any time in most years particularly in dry season.

The chroma is high is some deeper horizons. This is the basis for definition of aeric sub-group. Soils having thin layers of fine pyroclastic materials near the surface are excluded from the typic sub-group because such layers indicate an intergrade sub-group to Andaquepts. A dark Ap horizon or a moderately thick, dark A1 horizon are bases for defining intergrades for Humaquepts to Aquolls, according to base status. A lithic contact within 50 cm. is the basis for definition of lithic sub-group, a convention used throughout the soil taxonomy. Deep wide cracks in the expanding clays indicate intergrades to vertisol. *Jarosite* mottles of associated with low pH value indicate intergrades to sulfaquepts :

So Typic Haplaquepts are Haplaquepts which have —

(a) 50 per cent or more in the mass between the A1 or AP and 75 cm or more of the following :

 (i) If mottled and if hues are 2. 5Y or redder and moist values are 5 or more, moist chromas are 2 or less.
 (ii) If mottled and if hues are yellower than 2.5Y moist chromas are 2 or less.
 (iii) If with or without mottles, moist chromas are 1 or less;
 (iv) hues are bluer than 5Y.

(b) lack a layer in the upper 75 cm that has a texture finer than loamy fine sand;
(c) have organic matter content that decreases regularly with depth and unless a lithic or a paralithic contact occurs at shallower depths, reaches levels of 0.35 per cent or less within 1.25 m of the surface.
(d) lack a buried Histosol within 50 cm of the surface.
(e) have a A1 horizon of less than 10 cm thick that meets the colour requirements of Umbric or a Mollic epipedon; if cultivated have an Ap horizon with a moist value of 4 or more and a dry value or 6 or more;
(f) have ' N' value of less than 0.9 between 50 and 80 cm and less than 0.5 in all layers between 30 and 50 cm.
(g) lack a lithic contact within 50 cm of the surface.
(h) lack the following combination of characteristics

 (i) Cracks at some season in most years that are 1 cm or more wide at a depth of 50 cm
 (ii) More than 35 per cent clay in all horizons between 25 cm and either 1 meter or a lithic contact that is between 50 cm and 1 meter.

Typic Haplaquepts have the following families :

(1) coarse loamy
   (a) shallow on sandy skeletal
   (b) very deep

(2) coarse silty
   (a) very deep on sandy skeletal

(3) Fine silty
   (a) very deep

Table 6.10 : Profile Character of Coarse Silty, Typic Haplaquepts at Chidrawala (Auger Horing result)

Landform : Jakhan — Song interfluve
Relief : almost flat and level
Land use : Agricultural land (Recent)

Depth (in cm)	Characteristics
0 - 12.5	Gray (5Y5/1) and common rusty specs (7.5YR/6) silt loamy film and sticky, non-acids, chromas, of more than 60%)
12 - 25	Grayish brown (2.5 Y 5/2) and common iron mottlings (7.5 YR 5/6); silt loam, firm on moist and sticky on wet non-acid, chromas of 2, more than 60 per cent.
50.5 - 62.5	grayish brown (2.5 Y 5/2) common iron mottling (10 YR 3/4), loam, firm on moist and sticky on wet, non-acid chromas of 2, more than 60 per cent.
62.5 - 100	Grayish brown (2.5 Y 5/2) common iron mottlings (10 YR 3/4, 4/4) loam firm on moist and sticky on wet, non-acid, chromas of 2 or less 60 per cent
100 - 112.5	dark grayish brown to brown (10 YR 4/2.5) reduction (10 YR 6/1) mottling, loam, firm moist and sticky on wet, few gravels, non-acid chromas of 1 less than 20 per cent.

Source : Prepared from boring results of field survey.

Aeric Haplaquepts, a sub-group of Haplaquepts that are similar to Typic Haplaquepts except for (a) or (a) and (e)

The Aeric Haplaquepts have also been subdivided into following families :

(1) Coarse loamy
   (i) deep on sandy skeletal
   (ii) very deep
(2) Coarse silty
   (i) deep or sandy skeletal
   (ii) very deep

Table 6.11 : Profile Character of Aeric Haplaquepts at Chidrawala

Location : Chidrawala near the Song river
Loadform : Jakhan — Song interfluve
Relief : Almost flat and level
Land use : Agriculture

Depth (in cm)	Characteristics
0 - 25	Colour 10 YR 4/2, silt loam, friable, on moist and slightly sticky on wet, non-acid
25 - 37.5	Colour 10 YR 4/2, silt loam, few faint iron mottles (7.5 YR 4/4) few coarse reduction mottle (10YR 5/1), slightly firm on moist and moderately sticky on wet non-acid

(Contd.)

Table 6.11(Contd.)

37.5 - 75	Colour 10 YR 4/3 to 5/3 silt loam, common faint iron mottles (7.5 YR 4/4) and (10 YR 5/6) and reduction mottles (10 YR 5/1) slightly firm on moist and moderately firm on moist and moderately sticky on wet, non-acid.
87.5 - 100	Colour 2.5 Y5/2, silt loam, common iron mottles (10 YR 5/4) slightly firm on moist and moderately sticky on wet, non-acid

Source : Prepared from boring results of field survey.

Mollic Haplaquept

Another sub-group of Haplaquepts that are similar Typic Haplaquepts except for (e) and the A1 or AP meets all the requirements for a mollic epi-pedon except thickness.

The Mollic Haplaquepts have further been sub-divided into the following families which are identified in this study region.

(1) coarse loamy
deep on sandy skeletal

(2) coarse silty
deep on sandy

Table 6.12 : Profile Character of Coarse Silty Mollic Haplaquept near Rishikesh

Landform : Back water swamp
Relief : Slightly concave
Land use : Waste land

Depth (in cm)	Characteristics
0 - 8/10	black - (10 YR2/1), silt loam, fine granular (top 0.5 cm) strong medium to coarse angular blocky to weak prism like, firm on moist, moderately sticky on wet, few fine and coarse pores, slight biological activities, non-acid abrupt an slightly curved boundary.
10 - 40	Dark gray (10 YR 4/1, 80%) distinct common iron mottling (7.5 YR 5/5) fine and medium weak subangular blocky, firm on moist and moderately sticky on wet, many coarse and fine pores, moderate biological activities, non-acid, abrupt and slightly curved boundary.
40 - 53	Dark gray (10 YR 4/1, 60%) faint many mottlings. The colour of the mottling is 10 YR 3/4, sand with a thin band of silt layer massive, loose on moist and now sticky on wet, many coarse pores, slight biological activities, few medium and fine gravels non-acid, abrupt and slightly curved boundary.
53 - 80	Gray to dark gray (10 YR 4.5, 5/1 70%) many faint mottlings (10 YR 4/3), sandy loam, massive to coarse sub-angular blocky, slightly firm on moist and slightly sticky on wet common coarse pores, slight biological activities, micaceous, non-acid.

Source : Prepared from boring results of field survey.

Ochrepts

Ochrepts are another sub-order of the Inceptisols. The following observations are made and it is brought to the conviction that under mentioned standard criteria have established its identity :

(1) An ochric epi-pedon, or an umbric epi-pedon, no more than 25 cm thick if (a) the mean annual soil temperature is 8.3°C or more or (b) the mean summer soil temperature is more than 15. 5° at 50 cm depth if there is no horizon.

(2) 5°C or more difference between mean summer and winter soil temperature at 50 cm depth or at a lithic or a paralithic contact whichever is shallower.

(3) Chromas too high or hues too red for Aquepts or are never saturated with water and are not artificially drained.

One great group of Ochrepts occurs in this area. They are Ustochrepts.

The Ustochrepts are Ochrepts which

(1) are dry for 90 cumulative days or more in most years in some sub-horizon of the soil between 25 cm and 1 meter or a lithic or a paralithic contact whichever is shallower, or between 5 cm and a lithic or paralithic contact shallower than 50 cm; but are not continuously dry in all parts of the soil between these depths, for as long as 60 consecutive days unless the mean annual soil temperature is 22°C or higher.
(2) lack of a fragipan or a duripan.
(3) have a mean annual soil temperature of 8.3° C or more or have a mean summer soil temperature at 50cm or at a lithic or a paralithic contact, whichever is shallower, of either 15.5°C or more of cultivated or without an O horizon or 8.3° C or more if with O horizon.

In this area Typic Ustochrept which is a sub-group of Ustochrept is found. Typic Ustochrepts are soils with the following criteria :

(a) lack a layer in the upper 75 cm that has a texture finer than loamy sand, that is as much a 25 cm thick.
(b) lack mottles with chromas of 2 or less within 75 cm of the surface.
(c) have base saturation by (NH_4OH) of more than 60 per cent in some part of the soil within 75 cm of the surface.
(d) have organic matter content that decreases regularly with depth and unless a lithic or a paralithic contact occurs at shallower depth, reaches levels of 0.35 per cent or less within 1.25 m of the surface.
(e) lack a lithic contact within 50 cm of the surface.
(f) lack the following combination of characteristics
    (i) cracks at some season in most years that are 1 cm or more wide at a depth of 50cm and that are at least 30 cm long in some part.
    (ii) more than 30 per cent clay in all horizons between 25 cm and either 1 meter or a lithic contact that lies between 50 cm and 1 meter.
(g) have less than 40 per cent carbonate below the cambic horizon to a depth of 1 meter.

Typic Ustchrepts are non-acid and have only one family namely coarse loamy, very deep coarse loamy Typic Ustochrepts are very deep and occur in the area under investigation.

Table 6.13 : Profile Characteristics of Typic Ustochrept at Rishikesh

Landform : Ganga terrace
Relief : Almost flat and level
Land use : Partly urban and partly under cultivation

Depth (in cm)	Characteristics
0-25	Very dark greyish brown (10 YR 3/2), sandy loam, massive friable on moist and non-sticky on wet, common excremets, micaflakes, non-acid, diffused boundary.
25 - 40/45	Dark brown (10 YR 3/3), sandy loam, moderately fine and sub-angular blocky, friable on moist and non-sticky on wet, common excrements common medium and fine and few coarse pores and micaflakes, non-acid diffuse curved boundary.
45/45 - 76/84	Dark yellowish brown [10 YR 3/4), sandy loam, moderately medium to coarse sub-angular or blocky, many excrements many medium and fine but few coarse pores, micaflakes, non-acid, diffused and curved boundary.

(Contd.)

Table 6.13(Contd.)

76/84 - 120	Dark yellowish brown (10 YR 3/4), sandy loam, weak medium and coarse sub-angular blocky, many medium and fine and few coarse pores, micaflakes, non-acid

Source : Prepared from boring results of field survey.

Mollisols

The high piedmont level of the study area represents Mollisols which are characterised by mineral soils, having either mollic epi-pedon or having a surface horizon that are making to 18 cm, meets all the requirements for a mollic epi-pedon. In Mollisols upper sub-horizon is an argillic or natric horizon, that is separated from surface horizon by an albic horizon, that is more than 7.5 cm thick and that meets the colour organic matter, base saturation and structure requirements of a mollic epi-pedon, but do not have.

1. An oxic horizon.
2. A spodic horizon.
3. An argillic horizon.
4. At a depth of 18 cm have, as much as 30 per cent clay in all horizons down to a lithic or paralithic contact, or to a calcic horizon of 1 meter whichever is shallower; if there are cracks in some season in most years at least 1 cm wide at a depth of 50 cm. There is slicken sides close enough to interact at some depth between 25 cm and 1 meter or wedge-shaped or parallel shaped natural agregates with long axis tilted to 10° to 60° from horizontal between 25 cm and 1 meter.

Mollisols of this region have been sub-divided into the following sub-orders :

1. *Ustolls*

Ustolls are Mollisols which —

(1) have temperature higher than 8.3°C or higher chromas.
(2) have no albic horizons with characteristics associated with wetness, namely mottles, iron-manganese concretions larger than 2 mm or both.
(3) are not saturated with water at any season or the characteristics associated with wetness defined for aquolls.
(4) have no calcareous horizon except a calcic horizon that immediately underlies the mollic epi-pedon at depth of less than 50 cm and that has no more 40 per cent calcium carbonate equivalent.
(5) have one or more of the following combinations of properties.

(a) are dry for 90 cumulative days or more in moist years in some sub-horizons of the soil between 25 cm and 1 meter or a lithic or paralithic contact, whichever is shallower or between 5 cm and a lithic or paralithic contact shallower than 50 cm and are not continuously dry in all parts of the soil between these depths for as long as 60 consecutive days in more than 7 out of 10 years. Unless the mean annual soil temperature is 22°C or higher or the mean summer and winter soil temperatures differ by less than 5°C or both.
(b) If there is a lithic or a paralithic contact within 50 cm of the surface have base saturation (by NH_4OH) of 80 per cent or more in all parts above the contact.
(c) have either a calcic horizon or a horizon with concentration of soft powdering lime within 1.5 m of soil surface or within 50 cm below the base of any cambic or argillic horizon.

Ustolls have been identified in this region as Haplustolls, a great group of Ustolls which have —

(1) no argillic or natric horizon.
(2) no duripan with its upper boundary with in 1 m of the surface.
(3) no calcic or gypsic horizon with its upper houndary within 1 m of the surface unless some part of horizon overlying the calcic horizon is free of carbonates after mixing the upper 18 cm and has a texture of loamy fine sand.
(4) a transition between the mollic epi-pedon and the underlying horizon that has less than 25 per cent by volume of worm holes, worm casts or filled animal burrows.

Haplustolls have been further identified as *Udic Haplustolls,* a sub-group of Haplustolls which—

(a) lack mottles with chromas of 2 or less within 1 m of the surface or are not continuously saturated with water for 90 days within 1 m, of the surface.
(b) have a mollic epi-pedon less than 50 cm thick.
(c) lack within 1 m of the surface any brittled horizon, 15 cm or more thick; that contains some opal coatings or some durinodes.
(d) have a mollic epi-pedon overlying a cambic horizon, or have a molic epi-pedon that is non-calcareous, at least in the upper half. Its lower portion has one or more of the following :
    (i) moderate or strong prismatic, blocky or sub-angular blocky structure.
    (ii) higher chroma than the upper portion.
    (iii) redder hues than the upper portion.
(e) have a regular decrease in organic matter content with depth to a level of 0.5 per cent (0.29 per cent organic carbon) or less within 1.25 meter of the surface .
(f) lack lithic or paralithic contact within 50 cm of the surface.
(g) lack of petrocalcic horizon within 1 meter of the surface.
(h) lack a salic horizon within 75 m of the surface.
(i) lack either.
    (i) a calcic horizon of soft powdery secondary lime within 50 cm of the surface if texture is loamy, 75 cm of sandy and 45 cm if clayey or;
    (ii) If carbonates absent, some sub-horizon between 25 cm and 1 meter, or above a lithic or paralithic contact, is dry more than 135 cumulative days in most years of the mean annual soil temperature is 15°C or higher.
(j) lack the following characteristics —
    (i) Cracks at some season in most years that are 1 cm or more wide at a depth of 50 cm and that are at least 30 cm long in some part.

Udic Haplustolls have following 'family' namely —

(i) Coarse loamy non-acid
    (a) shallow on sandy skeletal
    (b) deep on sandy sketetal
    (c) very deep
(ii) fine silty
    (a) very deep

Table 6. 14 : Profile Characteristics of Deep Fine Loamy Udic Haplustoll at Bibiwala

Landform : Piedmont
Relief : nearly level and flat
Land use : forest with thin undergrowth of seedlings

Depth (in cm)	Characteristics
0 - 2	Very dark greyish brown (10 YR 3/2), sandy loam, strong fine crumb to strong fine sub-angular blocky, friable on moist and non-sticky on wet, strong biological activities, non-acid, abrupt and smooth boundary.
2 - 6	Very dark grayish brown (10 YR 3/2), sandy loam dark grayish brown to grayish brown (10 YR 4.5/2) platy, hard on dry, friable on moist and slightly on wet. Common medium and fine pores, common excrements non-acid abrupt and smooth boundary.
6 - 15	Very dark grayish brown (10 YR 3/2), sandy loam, moderately coarse and very coarse, sub-angular blocky, slightly hard on dry, slightly firm on moist and slightly sticky on wet, many medium fine pores, many excrements, non-acid, gradual and smooth boundary.
15 - 31	Very dark grayish brown (10 YR 3/2) sandy clay loam, weak medium and coarse sub-angular blocky firm on moist and slightly sticky on wet, common coarse and many medium and fine pores, very any excrements, non-acid, diffused and smooth boundary.
31 - 46/51	Brown (7.5 YR 4/4) and with dark brown humus coating (7.5 YR 4/2), sandy loam or sandy clay loam, weak medium coarse sub-angular blocky, friable on moist and slightly sticky on wet, many medium and fine few coarse pores, many excrements, non-acid, diffused and smooth boundary.
46/51 - 74	Brown (7.5 YR 4/4) with dark brown humus coating (7.5 YR 3/2); sandy loam or sandy clay loam, moderate fine to medium sub-angular blocky, friable on moist and slightly sticky on wet, many medium and fine and few coarse pores, many excrements non-acid, diffused and smooth boundary.
74 - 80	Brown (7.5 YR 4/4) with humus coating (7.5 YR 4/3) sandy clay loam, moderate very fine, sub-angular blocky, friable on moist and slightly sticky on wet, many medium and fine and few coarse pores, very many excrement, non-acid, abrupt and smooth boundary.
84 - 90	Brown (7.5 YR 4/4) sandy caly loam with coarse and fine gravels (50%), few roots horizontals.

Source : Prepared from boring results of field survey.

Aquic Udic Haplustolls are another sub-group of haplustolls that have criteria smaller to udic haplustolls except for Aquic haplustolls occurring in the lower slope of the high piedmont level.

Aquic Haplustolls have only one family as —

(1) Fine silty, non-acid
very deep, eroded phase

Table 6.15 : Profile Characteristics of the Fine Silty Aquic Udic Haplustolls Near Gulathappar Forest Rest House

Landform : Piedmonst high levels
Relief : nearly flat on top and sloping at edges
Land use : Forest

Depth (in cm)	Characteristics
0 - 12.5	Dark brown (10 YR 5/5); silt loam, slightly firm on moist and sticky on wet, non-acid.
12.5 - 37.5	Brown (10 YR 4/3) silty clay loam, moderately firm on moist, sticky on wet, non-acidic soil.
37.5 - 75	Yellowish brown (10 YR 5/5), common faint mottled with light brownish gray (10 YR/2), common salt, clay loam, and sticky, non-acid.
75 - 125	Yellowish brown (10 YR 5/5) common distinct mottled with (10 YR 6/1), light gray, non-acid.

Source : Prepared from boring results of field survey.

7

PHOTO-GEOGRAPHY OF SOIL DISTRIBUTION AND SOIL LEGENDS OF RISHIKESH AREA

Soil study of the area is carried out with the help of aerial photographs and satellite imagery of the region. This is followed by field work of the sample area and further, it has been interpolated and extrapolated by the knowledge and experience of local people and the kind of parent materials on which the soils are developed. In this area soils are of both coarse and fine textural types, indicating differences in kinds of parent materials deposited by the rivers coming from different source areas of the Himalaya and the Siwaliks. The maturity and degree of development of the soil profile reveal the comparative age of deposition, as well as the rate of soil of the high piedmont levels. The soils of the area are more mature than the soils of the Jakhan-Song interfluve. Within the Jakhan-Song interfluve the soils of the *bhabar* are immature as compared to the soils of the *tarai*. The later reveals a rapid rate of soil formation under extreme moisture conditions. Similarly, the high level of the alluvial fan area has less mature soils. The stages of development of soil profiles are correlated to the ages and levels of the different terraces, provided that similar conditions prevail on these.

Dissimilar conditions will result in dissimilar soils as in the case between depression of silted up river beds and slip-off slopes on the terraces.

Soils of High Piedmont Level

In the high piedmont level, the following soil families are developed—

1. Udic-Haplustolls
   - (i) Coarse loamy,
     deep on sandy skeletal.
   - (ii) Fine loamy
     - (a) deep on sandy skeletal.
     - (b) fine loamy,
       deep on sandy skeletal; and
       very deep
   - (iii) Fine silty
     very deep
2. Aquic-Udic-Haplustolls
   - (i) Fine silty
     - (a) very deep

FIG. 7.1 : Geocoded Satellite Imagery of Coloured Band, 4, 5, and 7 Dehra Dun

Deep, Non-Acid, Coarse, Loamy Udic Haplustolls on Sandy Skeletal

These are well drained soils overlying sandy skeletal materials on very gently sloping to level land surface. The slope is 1 to 3 per cent. The surface is covered with thin under-growth of shrubs and seedling and generally a thin layer of partly decomposed leaf litters and low grasses . The colour of the surface soil is (10 YR 3/2) very dark grayish brown, indicating high content of organic matter which forms the mollic epi-pedon. The structure is fine crumb to sub-angular blocky. The surface soil is friable when moist and slightly sticky when wet. The sub-soil colour is dark brown (7.5 YR 4/4) indicating formation of cambic horizon. The sub-soil textures are sandy loam slightly heavier than surface indicating local clay formation. The consistency is friable and slightly sticky. The structure is weak, medium to coarse, sub-augular blocky. Many fine, medium and few coarse pores are present throughout the profiles which have medium to strong biological activities. The development is good but its penetration to deep sub-stratum is restricted due to presence of gravelly bed. As a result the forest does not grow very well. The soil is said to be moderately favourable for Sal, and probably also for other timber species. The soils are located by the right side of the Lalpani Rao Cho in association with fine silty very deep Udic-Haplustolls.

Deep to Very Deep, Non-Acid, Fine Loamy Udic Haplustolls

These are of similar in morphology except the texture of the sub-soils which are similar in morphology except the texture of the sub-soils which are of sandy clay loam type. The increase in clay percentage is due to formation of clay, locally in the sub-soil horizon, as the illuviation of clay is not recognised by the formation of clay . The colour of the sub-soil is redder (5 YR 4/4) than in the coarse loamy Udic Haplustolls. The water-holding capacity is probably higher. The root development is good. The soil can be interpreted as favourable for sal and probably for other timber species like sisham, haldu, etc. The soils are located near the Barkot-Bibiwala forest road and are under forest.

Very Deep, Non-Acid, Fine Silty Udic Haplustoll

They are also similar to fine loamy very deep Udic-Haplustolls except in texture of the profile. The surface soils are loam to silt loam and the sub-soil is silty clay loam. The consistence of the surface soil is friable on moist and moderately sticky on wet conditions. The root development is good. The water-holding capacity is higher than the fine loamy family. The soil is very favourable for sal and probably other timber species. In the region of piedmont at high level, the fine silty family covered approximately 60 per cent of the surveyed area. It is located either on sides of the Bangali Rao Cho and Gulathappar Rao Cho and adjacent to Chandna Rao river terraces which are situated at higher level.

Very Deep, Non-Acid, Silty Aquic Udic Haplustolls

These are moderate well drained to somewhat imperfectly drained soils on the lower slope of very gently sloping to level uplands, dissected by occasional gullies. The slope is 1 to 3 per cent on the flat top and about 5 to 15 per cent at the edges of the unit. The surface soil is subject to moderate to severe sheet erosion. Hence, that depth of the dark surface soil (10 YR 3/2) is shallower than 25 cm. The texture and structure of the surface and sub-soil are similar to fine silty Udic-Haplustolls. The sub-soil colour below the cambic horizon has chromas of 2 or less and values of 5 of higher. Faint to distinct iron and reduction mottles indicate the saturation with water during at least a part of the year within 1 meter of depth. This may affect the growth of the existing forest. The eroded sloping edges of the piedmont have gravelly phases. The soil is mapped in association with fine silty very deep Udic-Haplustolls. The latter occurs not only slightly eroded flat remnants of the original piedmont surface and will be moderately suitable for sal and probably other timber species, if soil conservative measures are taken up.

Soils of the Jakhan Song Interfluve

Bhabar Soils

Soils of the bhabar of the Jakhan Song interfluve lie between the Song and the Jakhan Rao Cho adjacent to the high piedmont plain. The soils are classified as follows :

(a) Typic Ustifluvents
  (i) Coarse silty,
    very deep
  (ii) Coarse silty and coarse loamy on anisotropic very deep
  (iii) Coarse loamy
    very deep
  (iv) Sandy,
    deep on anisotropic
(b) Aquic Ustifluvents
  (i) Coarse silty,
    very deep
(c) Typic Xeropsamment
  shallow on sandy skeletal
(d) Typic Aquipsamment
  very deep

All the above soils are Entisols which were formed by alluviation of fluvial deposits. They show stratification in different textural layers or in broken texture lamination.

Very Deep, Non-Acid, Coarse, Silty Type Ustifluvent

It exists longitudinally and almost parallel to the Song river near to its left bank. The soils are well drained. The texture is silt loam or very fine sandy loam throughout the profile up to the depth of 100 cm. The structure is either massive or sub-angular blocky. The lamination is often partly disturbed by the biological activities indicated by the presence of common occurrence of excrements and common medium and few coarse pores. The consistency of the soil layer is friable on moist and slightly sticky on wet. The colour of the surface is dark grayish brown (2.5 YR 4/2) or grayish brown (2.5 YR 5/2) and that of sub-surface becomes grayish brown to light brown. The alternate layers in the sub-soil give rise to alternate hues of 2.5Y and 10 YR with alternate chromas of 2 and 3 respectively, indicating the geo-genetic colour and immature soil profile development. Only few micaflakes are present throughout the profile. The growth of forests in this soil is better than the adjoining areas but it shows a poor growth in comparison to the forests of high piedmont levels. We must assume that higher organic matter content and better biological activity have given Piedmont soil a better fertility, moisture holding capacity and rootability, as well as their textures are very good. They are favourable for the growth of sal and other timber species.

Very Deep, Non-Acidic Coarse, Silty and Coarse Loamy Silty and Coarse Loam, Anisotropic Typic Ustifluvent

They are similar in morphological character to the coarse silty very deep Typic Ustifluvents, except for the presence of alternate coarse silty and coarse loamy layer throughout the profile. Some sandy layers also occur. The soils lie by the side of the very deep coarse silty very deep Ustifluvent though the strong anisotropy will probably hinder root development and capillary rise of the moisture.

Deep, Non-Acid, Sandy Typic Ultifluvent Anisotropic

Soils are very well drained but not excessive as rapid percolation is partly restricted by the heavier soil materials in the deep sub-soil. The surface and sub-surface texture is loamy sand or loamy fine sand. The structure is loose single grain in the top soil. The colour of sandy top soil is grayish brown and of the sub-surface grayish to very light olive brown (2.5 Y 5/2 to 5/3). Biological activity is limited. The sub-soil is massive and laminated. Micaflakes are present throughout the profile.

Water holding capacity and availability of plant nutrients are generally low. The area under this soil is almost barren with patches of thorny tree and shrubs. The soil is not suitable for sal and other timber trees. It is only suitable for mixed forest of shrubs.

Shallow, Non-Acid, Typic Xeropsamment on Sandy Skeletal

This soil occurs in and near shallow bed of the Jakhan Rao Cho running parallel to the stream coarse. The soil surface often shows gravel and bears patches of thorny tree (21.2 Y plus species). The soils are excessively drained due to presence of the sandy skeletal material at shallow depths. The colour of the surface soil is grayish brown (2.5 Y 5/2). The texture of the surface soil is loamy sand with gravels, the structure is single grained and porous. The consistence is loose on dry, very friable on moist and non-sticky on wet. The area under this soil is unsuitable for timber forest.

Very Deep, Non-Acidic Coarse, Loamy, Typic Ustifluvents

These soils are found by the side of the Jhakan Rao Cho, where its water reappears in a perennial stream with sluggish meandering course. The soils are well drained. The colour of the surface soils is dark grayish brown (10 YR 4/2). The texture is sandy loam throughout the profile. The consistence is friable and non-sticky when wet. Few mica flakes are present throughout the profile. The colour of the sub-soil is brown (10 YR 4/3) indicating a tendency of slight development. The consistence of the sub-soil is slightly wet. The soil is under agriculture. Irrigation facilities cannot be made available from the water of the Jakhan Rao Cho. Citrus fruits are planted and grow well. Wheat and paddy are the agricultural produce of this soil. The soils have been brought under recent cultivation after the settlement of Punjabi refugees. The area lies in the lower slope of the *bhabar* and it receives seepage water from the adjacent up-stream during monsoon. The soils are moderately suitable for agriculture if irrigation is provided.

The Aquic-Ustifluvent and Typic Aquipsamments

Very Deep, non-acid coarse silty Aquic Ustifluvent are moderately to somewhat imperfectly drained soil occur in the forest plantation site of the Chandi Block. The colour of the surface is grayish brown (2.5 Y 5/2) and that of sub-soil is brown to dark brown mottle with iron and reduction mottle. The texture is silt loam followed by silt and very fine sand loam in alternate layers. The structure of the surface is massive with strongly disturbed lamination. The consistence is slightly firm on moist and moderately sticky on wet. The sub-soil structure is either laminated or massive with strongly disturbed lamination. Biological activity is moderate throughout the profile, with common medium and few coarse pores. Lamination is distinct in the lighter textured materials. The iron mottles are distinct particularly in the laminated banding and cross beddings, indicating the lateral movement and hinder vertical movement of water. As a result in the extreme dry state, the surface and sub-surface soil may loose capillary contact, which prevents the plants from reaching the available moisture in the sub-soil. The sub-soils have common faint iron and reduction mottlings indicating seasonal water logging specially during the monsoon. This will interfere with the growth of the trees like sal.

The area under this soil was previously covered with throny bushes and is at present under Sissam plantation. The soil is moderately suitable for the timber species other than sal, which does not grow well in aquic conditions.

Deep and Very Deep, Non-Acid Typic Aquipsamments

These soils record sandy skeletal substrata, being characterised by loamy sand throughout the profile down to 50 cm. or deeper. They overlie sandy skeletal material. They have developed under waterlogging conditions resulting from the high ground water tables in the terai. They also occur in the northern part of the terai. The soil profile, though the texture is excessively drained becomes internally ill drained because of temporary rise of ground water. The surface is sandy loam or loamy sand. The texture of the sub-surface and subsoil is loamy sand. The structure is loose single grained throughout the profile. Faint iron and reduction mottles are present in the sub-soil indicating the gleying conditions due to saturation with water during a part of the year. The soils are loose when dry, very friable on moist and non-sticky on wet. Biological activity is low. The area under this soil exhibits poor growth of vegetation probably due to the extreme moisture regime at least for a part of the season.

Soils of the Jakhan-Song Interfluve

Soils of the *terai* of the Jakhan-Song interfluve are developed under either or at least temporarily extreme moisture condition which characterise the mottled cambic horizon. This indicates the formation of Inceptisols except in case of sandy texture which belongs to Entisols and sandy soils of recent formation is called Psamment. As a result the following soils are found in the *terai* region.

(a) Typic Haplaquepts
Coarse silty, deep on sandy skeletal and anisotropic coarse loamy, deep on sandy skeletal and anisotropic fine silty, very deep.

(b) Aeric Haplaquepts,
Coarse loamy, very deep.

(c) Aquic Ustifluvents,
coarse silty
deep on sandy skeletal
very deep.

(d) Typic Aquipsamments,
shallow on sandy skeletal.
deep on sandy skeletal
very deep.

(e) Typic Xeropsamments,
very deep.

Deep and Very Deep, Non-Acid, Coarse Silty Typic Haplaquepts

These soils are ill-drained, overlying the anisotropic or the sandy skeletal material. The colour of the soils is gray with a chroma of 2 or less and a value of 5 or more. The hue varies from 10 YR to 5Y. The sub-surface soil has dark yellowish brown to strong brown mottles with chroma of 4 or more and value of 3 or more in hues of l0YR to 7.5YR. The structure of the surface soil is weak, fine and medium sub-angular blocky and that sub-surface is fine and coarse sub-angular blocky. The consistence is firm on most and slightly sticky on wet, throughout the profile. Biological activity is moderate, the soils are non-acid. The area under this soil in the *terai* is under intensive cultivation. Paddy, wheat, sugarcane and other vegetables are grown here. The soil is very suitable for agricultural crop as good capillary raises the underground water level and keep the soil wet. Occasional irrigation is necessary for better result in the crop yields.

Deep and Very Deep Non-Acid Coarse Loamy Typic Haplaquepts

These soils are on sandy skeletal are similar in morphological character to the Typic Haplaquepts, coarse, silty, deep and very deep respectively, except for the texture which is sandy-loam throughout. The consistence is friable on moist and slightly sticky on wet. The area under this soil is also under intensive cultivation for argicultural crop. Time to time irrigation is necessary to obtain better yields.

Very Deep, Non-Acidic Coarse, Silty Aeric Haplaquepts

Soils of this type are ill drained, similar in morphological characteristic other than texture and consistence. The texture is silt clay loam to clay loam with depth and the consistence is more firm on moist and sticky on wet than in the coarse silty family. The surface is subject to cracks on drying. The water holding capacity is high and the soils become wet during the cropping period. The area under this soil is only suitable for paddy, which gives high yields.

Very Deep, Non-Acidic Coarse, Silty Aeric Haplaquepts

These soils are somewhat imperfectly drained soils overlying sandy skeletal materials. The surface is mixed with uniformly distributed root remains of crops. The colour of the surface soil is dark grayish brown 2.5Y 4/2 to l0 YR 4/2 and that of the sub-surface and sub-soil ranges from hue of l0YR or yellower with chromas of 4 or less within 100 cm. The sub-surface soils and sub-soils are mottled. The reduction colour is l0 YR 5/1 to 2.5Y 5/2 and the oxidation colour 7.5YR 4/4 to l0 YR 5/6. The colours of the reduction mottles in top soil do not fulfil the requirements of the Typic Haplaquepts. The texture of the surface and sub-surface is silt loam followed by fine sandy loam or silt loam at depths. The soils are non-acid slightly firm on moist and moderately sticky on wet. Water holding capacity is favourable. Moreover, due to the medium ground water table (155 cm) and mostly uniform texture there is a good capillary rise of water to feed the agricultural crop. The areas under this soil are under extensive cultivation of paddy, wheat sugarcane and vegetables and are very suitable for cultivation through occasional irrigation, which also is instrumental to better crop yields.

Very Deep, Non-Acidic Coarse Loamy Aeric Haplaquepts

This type of soil is similar in morphology to the very deep fine silt family except for texture and consistence. Moreover, the coarse loamy family is ill drained to somewhat imperfectly drained and has comparatively higher groundwater table. The texture is sandy loam throughout the depth and the consistence is more friable. The area under this soil is under intensive cultivation of paddy, wheat and sugarcane. The irrigation requirement, is very little and the external and internal drainage prevent these areas becoming permanent swamps. The soil is suitable for all types of crop production.

Deep and very Deep Non-Acid Coarse Silty Aquic Ustifluvents

These are the soils on sandy skeletal which have been described in the section of the Jakhan Song interfluve. The soils have more optimum moisture conditions for supporting plants and are suitable for all types of crops. Further irrigation facilities make these soils capable of producing good crops.

Shallow and Deep Non-Acid Aquipsamment

These are developed on sandy skeletal, having similar morphology. This is decribed as very deep Aquic Xeropsamment. The only difference is pronounced reduction in case of the deep Aquipsamment. Paddy is grown on this soil. ' The shallow Aquipsamment is also cultivated with wheat and paddy but the very poor quality of this soils results in poor quality of crops.

Very Deep Non-Acid Typic Xeropsamments

This type occurs in this area and is located around Chiddrawala village. The land type under this family seems to be the old levee of Jakhan Rao Cho. As such, they are similar to the leave soils of the Lalpani Rao River in the alluvial fan, which are also Typic Xeropsamments which are well drained soils. The texture is loamy sand throughout the profile. The structure of the layers is loose single grained to massive and laminated. The consistence is loose on dry, very friable on moist and non-sticky on wet conditions. The biological activity is moderate. The water holding capacity of the soil is very low and the soil is greatly susceptible to severe drought hampering the growth of crops. The soils are therefore not favourable for cultivation of crops.

Soils of the Alluvial Fan

Soils of the alluvial fans have been deposited by the Lalpani Rao Cho. The fan comprises of the following land types :

(1) levee

(2) back water swamp
(3) older terrace
(4) incised courses of the river

The soils developed on different land types are different from one another. The soils on the river levee belong to the Typic Xeropsamments of the following families .

(1) shallow on sandy skeletal
(2) deep on sandy skeletal
(3) very deep

Typic Xeropsamment

This type of soil occurs near Chiddrawala villags. The land type under this family seems to be old levees of the Jakhan Rao Cho in this fan areas. Typic Xeropsamments are very well drained soils, often overlying coarse loamy materials in the deep substrata below one meter. The texture is loamy sand, and liberation of colour is more exposed in the downward trends. The texture is loamy sand throughout the profile. The structure of the layers is loose on dry, very friable on moist and non-sticky on wet conditions. The biological activity is moderate. The water holding capacity of the soil is very low and the soil is greatly susceptible to severe drought that will hamper the growth of crops. The soils are therefore not favourable for cultivation.

Soils of the Backwatar Swamp

The soils of the backwater swamp are :

(1) Deep, non-acid, course, silty Mollic Haplaquept on anisotrophic and sandy skeletal.
(2) deep, non-acid, course, silty Typic Haplaquept on anisotrophic and sandy skeletal material.
(3) very deep non-acid course loamy Aeric Haplaquept.

As the area under these soils lies in a depression, it is swampy during most of the year. As such, the area is not suitable for agriculture. The profile characteristics of the above soils have already been described in the section on soils of *Terai* except for those of the Mollic Haplaquept. The description of the same is given below,

Deep non-acid coarse silty, Mollic Haplaquepts

There are ill drained soils overlying amisotrophic and sandy skeletal material. The surface is covered with aquatic plants and grasses. The colour of the surface is very dark gray (10YR 2/1) indicating mollic epi-pedon. The depth of the mollic epi-pedon is less than 25 cm. The texture of the surface and sub-soil is silt loam followed by sand and sandy loam downwards. The sandy layers in the profile contain few medium and fine gravel and thin lenses of silty materials. The structure of the surface soil is medium to coarse angular blocky with tendency towards fine prism like structure that indicates the deposition of the surface due to rapid drying. This may be due to interference of sand layers of shallower depth where capillery rise of water from the ground water ceases. The sub-surface and sub-soil structure is fine to coarse sub-angular blocky except in the sandy layers where the structure is loose single grained (structure less) . The profile has moderate to little biological activity which is not uniform throughout the layers. Coarse and medium pores are common to many. The land under this soil is moist throughout the year. Cultivation is only possible if drainage is improved.

The soils of the incised river course of Lalpani Rao Cho are shallow, non-acid, coarse, loamy Mollic Haplaquept on sandy skeletal.

The shallow coarse loamy family differs from deep coarse loamy family in depth and in absence of formation of cambic horizon. Due to shallow depth, root penetration is restricted. The surface is covered with high grasses and thorny shrubs. The area is under Sisam plantation. The soil is poorly suitable for Sal and other timber species.

Soils of Terraces

Soils of the terraces belong to Entisols, Inceptisols, and Mollisols, in the ascending order of the terrace levels. As such most recent terrace have Entisols, whereas, the oldest terraces have Mollisols. The following soils are found on the terraces in the area under study with the help of aerial photographs.

1. *Typic Xeropsamments*
    - non-acid
    - micaceous
        - shallow on sandy skeletal
        - very deep
    - non-micaceous
        - shallow on sandy skeletal
        - calcareous (non-micaceous)
        - shallow on sandy skeletal
2. *Typic Ustifluvents (Non-acid)*
    - Coarse loamy
        - deep on sandy skeletal
        - very deep
3. *Typic Haplaquepts*
    - Coarse loamy
    - Deep or sandy skeletal
4. *Mollic Haplaquepts*
    - Coarse loamy
    - deep or sandy skeletal
5. *Typic Ustochrepts*
    - Coarse loamy
    - very deep
6. *Udic Haplustolls*
    - Coarse loamy
    - shallow on sandy skeletal
    - deep or sandy skeletal
    - very deep

The shallow non-acid non-micaceous Typic Xeropsamments on sandy skeletal materials, are found under forest and formed on the alluvium of the Song river, whereas;

The shallow calcareas non-micaceous Typic Xeropsamments on sandy skeletal lie, under active flood plain and contain calcium indicated by effervescense with HCl. The area is under forest. But the quality of vegetation is poor, probably due to occasional flooding.

Very deep, non-acid micaceous, Typic Xeropsamments occur in the lowest terrace level of the Ganga near Rishikesh. The soils are excessively well drained and do not contain sufficient moisture to raise a good crop . The soils are not suitable for agricultural crops and are utilized for grazing and for factory and building sites. Occasional cultivation of vegetables and orchards common near the bank of the Ganga. Some crops are raised by intensive use of irrigation.

Deep and very non-acid coarse loamy Typic Ustifluvents on sandy skeletal material on the Ganga terraces No. 2 and 3 and Chandna Rao terraces No. 2 and 3. The area of these soils are under cultivation. The soils may be suitable for agriculture only after irrigation.

Deep, non-acid, coarse loamy Typic Haplaquepts on sandy skeletal are found in older meander cut-off channels. The ground-water table is relatively higher than in the adjacent land of the same terraces. As such moisture conditions in the soils are very favourable for the growth of paddy. Sugarcane is also cultivated but it requires light irrigation during dry months.

Deep, non-acid, coarse loamy Mollic Haplaquepts on and sandy skeletal material — This type occurs in the swamps of old meander cut off channels on the Ganga terraces No. 2 and 4. The lands under this soil are unsuitable for crops due to wetness.

Very deep, non-acid, coarse loamy Typic Ustochrepts are well drained soils on the oldest Ganga terrace. The texture of this soil is sandy loam throughout the soil profile. The structure of the surface soil is massive and that of sub-surface and sub-soils is moderate to weak, medium to coarse sub-angular blocky. The colour of the surface and sub-surface is very dark grayish brown (10 YR 3/4). The soil profile has moderate biological activity. Medium and fine pores are present throughout the soil, associated with few to common coarse pores. The soils are friable on moist and slightly sticky on wet conditions. Though the colour of the surface is very dark grayish brown, it may not contain the organic matter sufficient for the requirement of mollic epi-pedon. The land under this soil is free from vegetation to satisfy the required organic matter content. The terrace is partly under urban development and partly under horticulture. If sufficient irrigation is provided the soil is suitable for agriculture.

Shallow and deep non-acid, coarse, loam Udic Haplustolls on sandy skeletal — This type occurs on the oldest Ganga and the Chandna Rao terrace. The very deep coarse loamy family occurs in the older terraces of Lalpani Rao Cho in the alluvial fan.

The difference between the very deep Udic Haplustolls on piedmont and older terraces is the differences in hue in the cambic horizon. The colour of the cambic horizon of this family on piedmont is redder than that on oldest terraces. This may indicate the difference in period of deposition of the parent materials on the two different land types. The soils are partly covered with low mixed trees and shrubs, partly with grasses. The forest plantation has been started on the coarse loamy Udic Haplustolls by the Forest Department of Uttar Pradesh Government. Part of the land under this soil near Rishikesh has been utilized for extension of the urban area.

The soil Map of Rishikesh (Fig. 7.2) has been prepared according to following legend based on 7th Approximation Soil Classification and Soil Taxonomy (*Ref.* Basic System of Soil Classification for Making and Interpreting Soil Survey) (USDA 1975),

Tentative Soil Legend

Non-Acid
Non-micaceous

1. Shallow on Sandy Skeletal
2. Deep on Sandy Skeletal
3. Very deep
4. Association of soils 1, 2 and 3.

Micaceous

5. Very deep

Calcareous (Non-micaceous)

6. Deep on sandy skeletal

Typic Aquipsamments (Non-acid; Non-micaceous)

7. Shallow on sandy skeletal
8. Very deep
9. Association of soils 7 and 8

Typic Ustifluvents (Non-acid)
Coarse loamy

10. Shallow on sandy skeletal
11. deep on sandy skeletal
12. Very deep
13. Association of soils 10 and 11
14. Association of soils 11 and 12 Coarse silty
15. Deep on sandy skeletal
16. Very deep

Coarse silty and coarse loamy anisotropic sandy

17. Very deep sandy
18. Deep on anisotropic

Association of Typic Xeropsamments and Typic Ustifulvents Non-acid)
Coarse loamy

20. deep on sandy skeletal
21. very deep
22. association of soils 20 and 21

Coarse Silty

23. deep on sandy skeletal
24. very deep
25. association of soils 23 and 24

Sandy

26. deep on anisotropic

Association of Typic and Aquic Ustifluvents (Non-acid)

27. association of soils of 17, 21 and 23
28. association of soils of 17 and 23

Association of Typic Xeropsamments, Typic Aquipsamments and Aquic Ustifluvents (Non-acid)

29. association of soils l, 8 and 26
30. association of soils 2, 10 and 12

Aeric Haplaquepts (Non-acid)

Coarse loamy

31. deep on sandy skeletal
32. very deep

Coarse silty

33. very deep

Coarse loamy

34. Shallow on Anistropic and Sandy skeletal

Coarse silty

35. deep on anisotropic and sandy skeletal

Fine silty

36. very deep

Association of Aeric and Typic Haplaquepts (Non-acid)

37. association of soils of 31, 32 and 35
38. association of soils of 31 and 35
39. association of soils of 33 and 36

Mollic Haplaquept (Non-acid)

Coarse loamy

40. deep on anisotropic and sandy skeletal

Coarse silty

41. deep on anisotropic and sandy skeletal

Association of Typic and Mollic Haplaquepts (Non-acid)

42. association of soils of 34 and 41

Typic Ustocrepts (Non-acid)

Coarse loamy

43. very deep

Udic Haplustolls (Non-acid)

Coarse loamy

44. Shallow on sandy skeletal
45. deep on sandy skeletal
46. Very deep

Fine loamy

47. Very deep

Fine silty

48. very deep

Associations

49. association of soils of 44 and 45
50. association of soils of 45 and 48
51. association of soils of 47 and 48

Aquic Udic-Haplustolls (Non-acid)
Fine silty

52. very deep

Association of Mollic Haplaquepts and Udic Haplustolls (Non-acid)

53. association of soils of 40 and 45

Association fluvents and Udic Haplustolls (Non-acid)

54. association of soils of 11, 15 and 47

Association of Udic Aquic Udic Haplustolls

55. association of soils of 48 and 52

Miscellaneous Land Types

56. Urban areas, factories etc.
57. Excavated site
58. Sandy sketal, active flood plain
59. Steep eroded terraces, valley

Phases of Soil Family

e_1 - Slight to moderate erosion
e_2 - Moderate to severe erosion

8

CASE STUDIES OF THREE TOWNS OF HIRAKUD BASIN

Introduction

Photo-geographical Analysis of the second selected region is attempted for the Hirakud Basin situated within the Mahanadi River Basin. Landuse of Mahanadi Basin which includes the Hirakud Canal Command Area is studied with the help of aerial photographs. Approximately 400 aerial photographs in the scale of 1: 50, 000 have been studied. The interpretation area covered is 11,100 sq. km. The following landuse categories has been dechiphered :

		sq. km.
(1)	Towns and cities	123.70
(2)	Villages	41.50
(3)	Orchard	5.50
(4)	Cropland	6,804.30
(5)	Scrub	65.00
(6)	Dense forest	1,800.00
(7)	Open forest	1,200.00
(8)	Water feature	750.00
(9)	Unproductive land	100.00
(10)	Wasteland	98.00
(11)	Grazing land	112.00
		11,100.00

It has been observed that cropland occupies more than 50 per cent of the land being followed by forests.

It has been found that the Canal Command Area falls under the District of Sambalpur in Orissa. Land which was dusty brown previously, now looks lush green even in the month of April. Due to Hirakud Canal irrigation, rice is grown throughout the year. As such the land appears green in most of the year. Further it has been observed that due to construction of Dam two new towns have emerged near the main dam. These two new towns are studied with the help of aerial photographs and Satellite Imagery. For in-depth study, a thorough field investigation is made and the morpho-analysis attempted in this chapter.

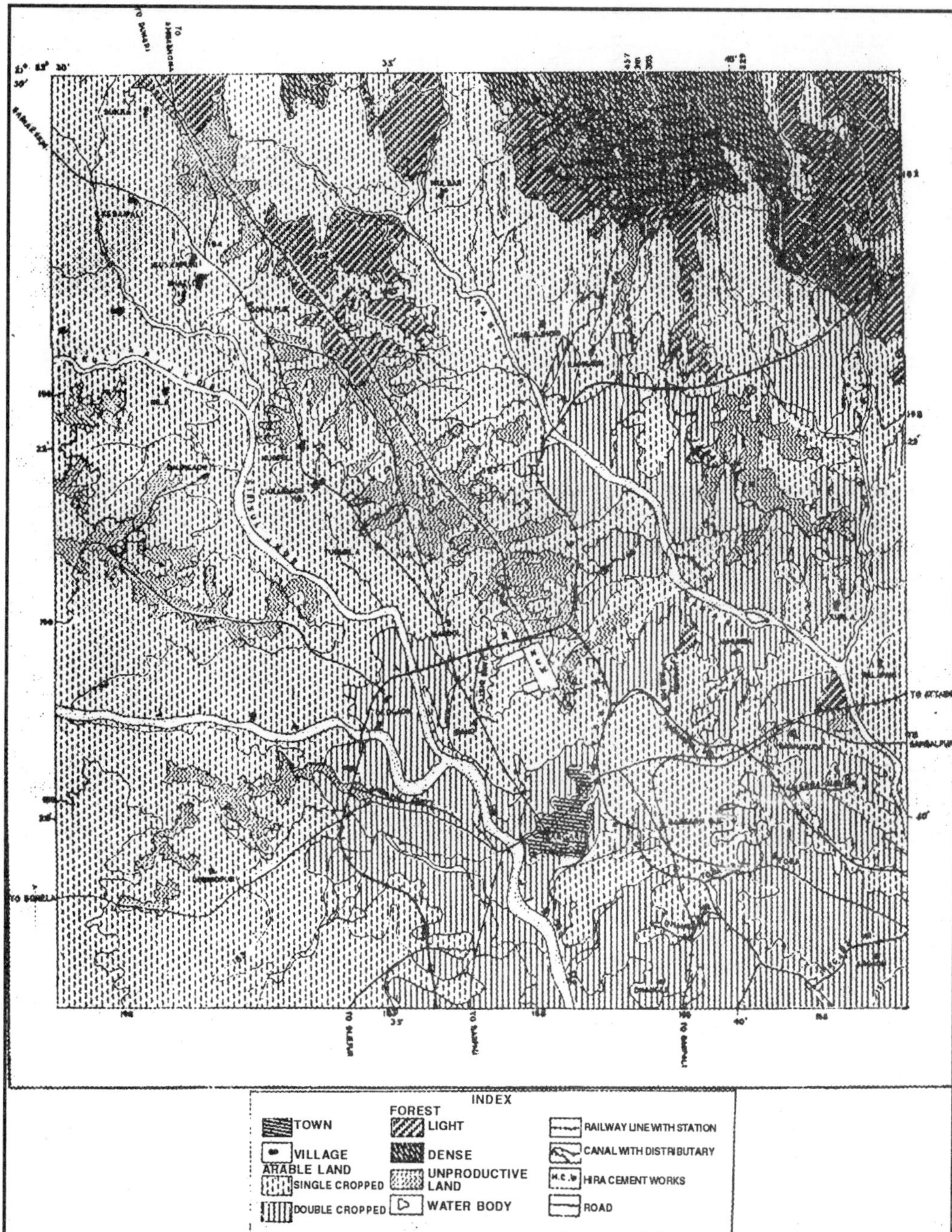

FIG. 8.1 : *Landuse Map of Hirakud Basin Orissa*

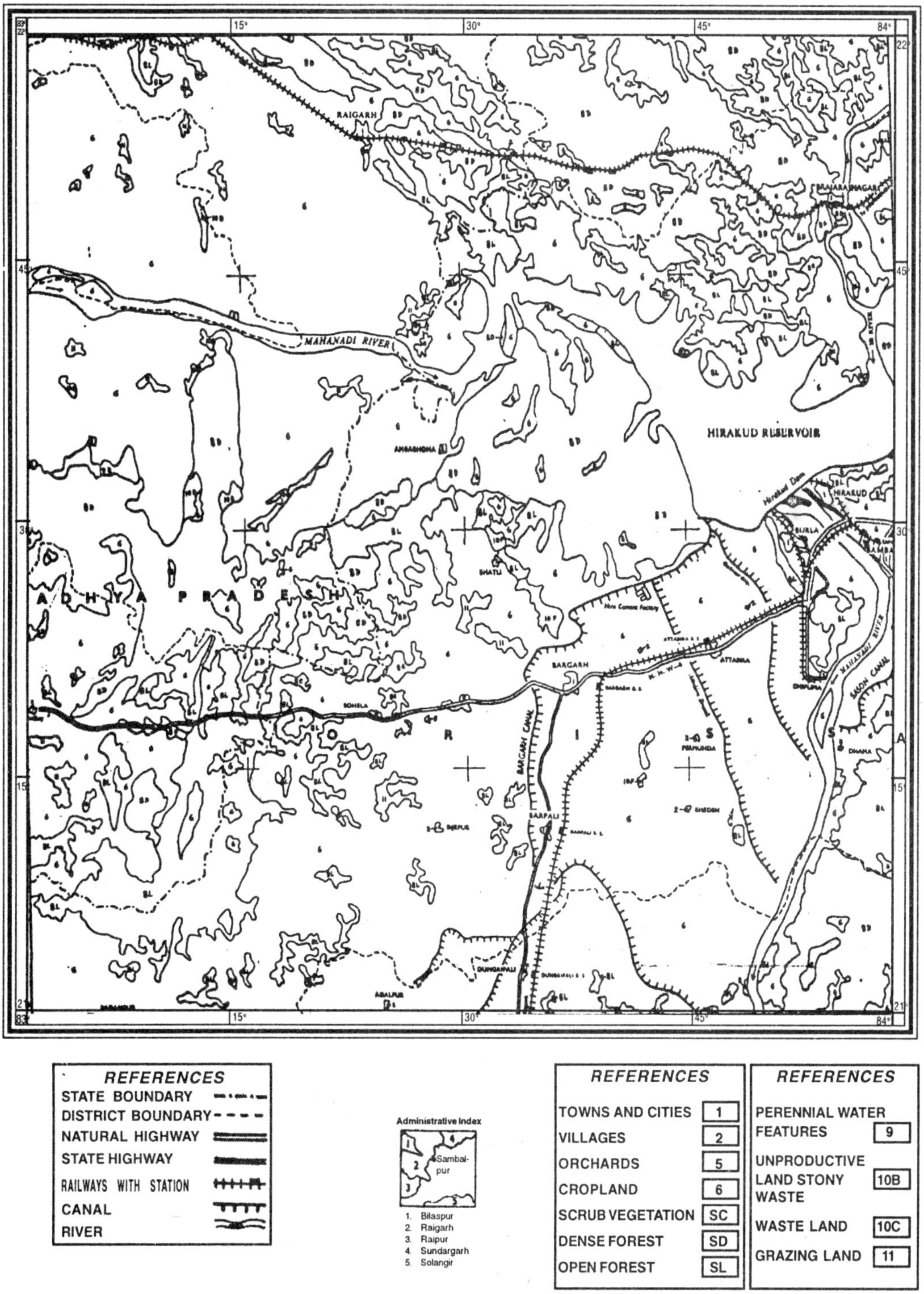

FIG. 8.2 : *Landuse Map of Mahanadi Basin Orissa and Madhya Pradesh*

Bargarh is another town which falls within the realm of this canal command area. The town is an old one, but gained importance due to Hirakud canal as trade and economy have been improved considerably. Besides the study with the help of aerial photographs, considerable field investigations have been carried in the area. In this chaptar three towns are described fully to understand the growth of development of the area after the introduction of the Hirakud Canal Command Area. These three towns are (1) Bargarh; (2) Burla, and (3) Hirakud.

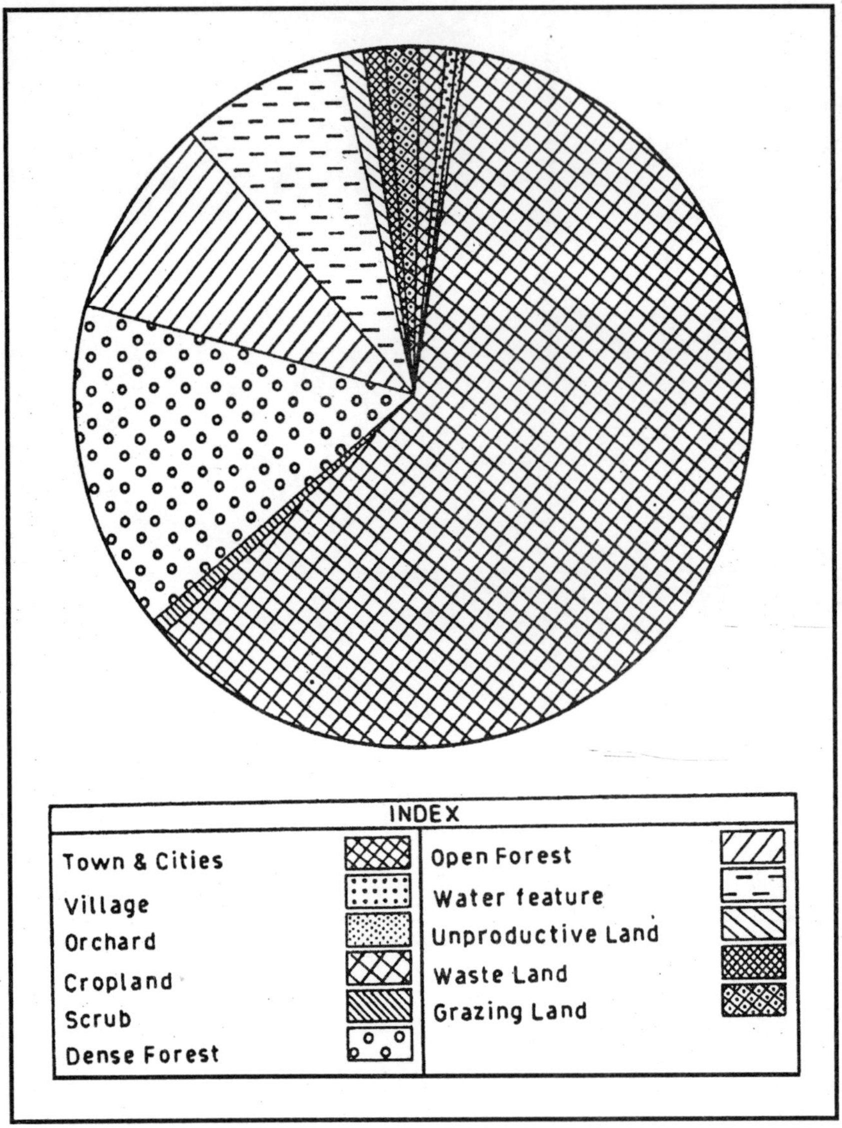

FIG. 8.3 : *Landuse Pattern in Mahanadi Basin*

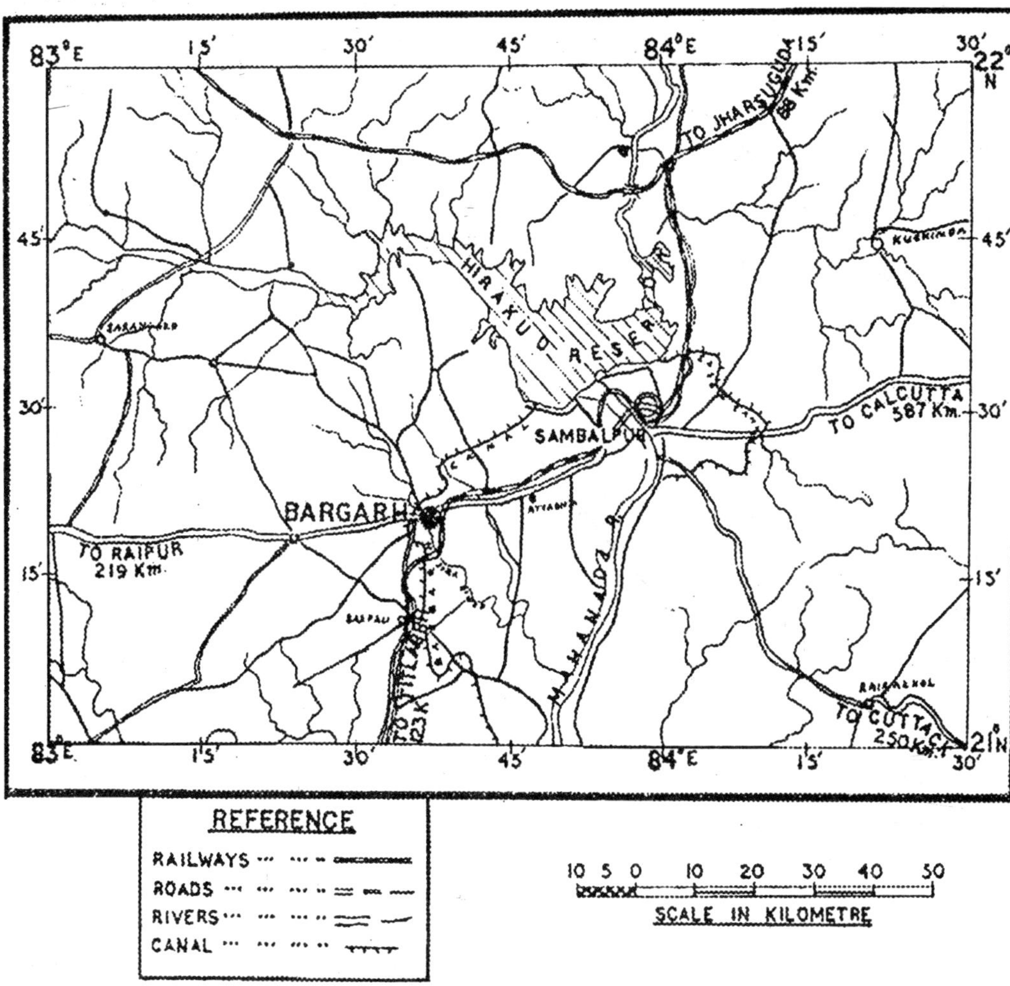

FIG. 8.4 : *Regional Setting of Bargarh Town*

BARGARH TOWN

Bargarh town is selected because part of it falls under irrigated area and part falls under non-irrigated area. The scenario is completely changed between the canal command area and non canal command area. Bargarh town is located at the edge of the Bargarh cannal: Two big canals which commands the irrigated area, namely (1) Bargarh canal and (2) Sasan canal.

This town is chosen due to its location on the ayacut of Hirakud Canal System covering an area of 13.21 sq.km. It is the headquarter of Bargarh sub-division of Sambalpur District of Orissa and lies at a distance of 59 km. west of the district headquarter and about 55 km. south-west of Hirakud Project site. It is served by S. E. Railway running from Jharsuguda to Titlagarh. Bargarh Railway Station is about 3 km. from the town. A meter gauge railway line connects the town with the limestone quarry at Dungri. This limestone is of Cuddapan age and the quality is of very high grade. The limestone are brought from here to the limestone factory which is situated in the town.

Calcutta-Bombay Highway (NH 6) runs through the heart of the town. To avoid congestion of traffic a diversion road is under construction. The main canal of the Hirakud Project-Bargarh canal passes west of the town.

The town has a municipality with a total population on 51,135 in 1991. The municipality was set-up in May 1952. Prior to this, there was a sanitation committee at Bargarh under the Central Province Village Sanitation Act 1902 and continued up to 1941 when a Union Board was formed here under the Sambalpur Local Self-Government Act 1939. The Union Board is now replaced by a Municipality.

The Municipality has been divided into 15 Wards against 14 in 1961 and 35 Blocks with 22 settlements nuclei most of which are locally known as 'Para'. They are as follows :

Ward No.	*Para*
1.	Giranadi
2.	Marwari para, Telipara
3.	Upar para, Kuli para, Das para
4.	Naik para, Upar para
5.	Satara para
6.	Marwari para
7.	Gondara, Mangua para
8.	Daktar Khan para, Gauda para
9.	Hat para
10.	Christian para
11.	Mohanti para
12.	Khajurlikra, Ghulipali para
13.	Gobindpali para
14.	Baltikra para

During Third Five Year Plan (1961-62 to 1965-66) the Town Planning Organisation of the State of Orissa selected six towns of the district of Samtialpur of which Bargarh is one for preparation of a Master Plan. But until recently Bargarh Master Plan is yet to be completed when the study is made.

Aerial photographs clearly reveal that the town lies on the Hirakud Basin of Upper Mahanadi Flood Plain, more prominently on the cliff side of the Jira River, the right bank tributary of the Mahanadi. The term 'Hirakud Basin', is coined after Hirakud town came into existence after construction of dam, popularly known as Hirakud Dam Circle. The town itself gained its importance as industries have grown up here. As Bargarh town forms a small part of Hirakud Basin, it exhibits a level plain characterised by thick deposits of alluvium. The local relief of the town ranges from 7.50 metre to 15 metre above sea level .

The town is under the influence of Indian Monsoons. Summer is generally hot and dry. Temperature begins to rise steadily from the begining of March. May is the hottest month with mean daily temperature of 41. 6°C (106°F) . Temperature begins to fall from the middle of November- December is the coldest month with 12.1°C (53.8°F) . Temperature sometimes goes down to 4°C by western disturbances. Average annual rainfall is 1500 mm. Most of the rainfall is received during monsoon months.

PLATE 1 : Bargarh Canal coming out of Hirakud Reservoir local people catching fish with hook and angles.

PLATE 2 : Close-up view of Bargarh Canal coming out of Hirakud Reservoir

Growth and Evolution

Few sythentic records are available regarding the origin of the town. So it is somewhat difficult to make an estimate of the town chronologically. However, based on some literature and information, collected during the field survey, three successive stages, can be traced regarding its origin and growth pattern. The original name of the place was Baghar Kota as known from an inscription of the llth century. The name Bargarh was coined probably from the time of Balaram Dev, the first Chauhan Raja of Sambalpur who made it for sometime his headquarter and constructed a big fort for protection. Narayan Singh the last Chauhan Raja, granted this place to two Brahmin borthers Khishna Das and Narayan Das, sons of Baluki Das who was killed in action by the Gond rebels led by Bandya Ray and Mahapatra Raj. The grant is popularly known as Sirkanta grant.

At the beginning of the present century, Bargarh was the western most 'Tahsil' of Sambalpur district with an area of 3,126 sq. miles. The 'Tahsil' contained 1,172 inhabited villages with Bargarh as the headquarter of the Tahsil.

Of course, geographical phenomena was no less active for its origin at that time. The meander loop of the Jira river attributed an attractive natural factor for its location and the first nucleus of the town had been formed out of the river bend of the vast natural landscape and had developed market administrative functions.

The second phase of its development can be traced from 1936 when Sambalpur was transferred to Orissa which became a separate province. The strategic importance of some nuclei around the original nucleus increased to provide its surroundings with economic, political and cultural facilities. As a result, the town began to rise very rapidly being patronised by the traders and sophisticated group of the rural areas as well as by the Government of Orissa by the attraction of its centrepetal force.

The real development of the town took place from the middle of the present century, particularly from 1947 when the construction of Hirakud dam was initiated. The artificially created reservoir on the bed of the Mahanadi about 19 km. upstream of Sambalpur submerged a number of inhabitated villages and thereby the rehabilitation of thousands of homeless people took place on the vast agricultural tract lying south of reservoir around Bargarh. In this manner the entire subdivision became fairly densely populated. Many cultural and material changes have taken place. New roads and other facilities have been created. Population also continues to pour in. These have helped the town to expand, imparting it a definite shape.

Development Process

The present chapter deals with something about the role of nuclei and cells in pattern formation and development factors of the town. The earliest nucleus was formed along both sides of NH 6 close to the loop of the Jira River. Later on administrative offices/commercial and educational institutions around the original nucleus led to the formation of this urban centre. The parallel ribs of roads within the town were formed by the intersection of the inter-movement between those administrative and commercial nuclei. Residential centres act as a stimulating force to its future development.

Though the town is of very ancient origin, its street pattern has been developed very recently. So the chequer-board street pattern is found in the developed part of the town. The converging point of roads from various directions has given birth to a central core of the town. Accordingly, various functions have developed along the central core.

Morphological Structure

In the developed part of the town several fractional zones of varying strength and intensity have been observed. The nature and functions of those zones are complicated and detailed though they appear distinctly on the map when plotted. They do not stand with any single performance. They have undergone and experienced changes in keeping pace with the phases of all round growth and development. Though the functional units of the town present complicated picture yet the town has given rise to several functional zones. They are as follows :

Residential

Residential areas of the town comprise of about 47 per cent of the developed areas and about 11.6 per cent of the total area of the town. In between the residential areas occur business and service sectors which are common features of the town. The ground floor of the double storeyed buildings along its main thoroughfare are mostly occupied by shops and offices. In some places it has been noticed that the back side of the office buildings are used for dwelling purposes. The transport oriented residential set-up is another notable feature of the town. High class residential buildings are found along the roads away from the congested part of the town to avoid unhygenic atmosphere. The middle class residential buildings are found mainly on the margin of the residential zones. The dwelling places of poorer classes are developed in the form of slums in different parts of the town. Most of them are on the lowlying agricultural tract, that is in the periphery of the developed area.

The following are the chief morphological characteristics of the existing residential areas of the town :

1. Influence of agricultural economy on the residential building status.
2. Aggregation of residential building based on caste, religion, occupation and segregation for social position, low income or for personal interest.
3. Three types of dwelling houses based on high, middle and low income groups.
4. Residential buildings are both *pucca* and *kuchcha*, double or multi-storied. Private residential houses are very few in number.
5. *Kuchcha* houses are characterised by earthen walls with thatched roof.

Commercial

The town has a very old commercial base. Since early times it has been functioning as a rural market centre. The Jira river offers an excellent business emporia. Before the inception of the grain and cattle market, its commerce was carried on through the collection of 'Kendu' leaves. It is well known that this area is famous for Kendu leaf production. Kendu leaves are required for ' bidi' production. The local Oriya and aboriginal people are engaged in this respect. They used to collect the leaves from the surrounding areas and sell it to the Sindi people who had much interest in it. Later, the Government of Orissa nationalised the Kendu leaf trade and took the initiative for its collection and distribution.

With the advent of Hirakud Project on the Mahanadi river, the roadways and railway linkages had much developed. This had led to quantitative and qualitative development of Bargarh's commercial function. Its role and responsibility as commercial centre increased with gradual development of other

PLATE 3 : View of the Concrete Dam of Hirakud Dam

PLATE 4 : View of Bargarh Canal, Hirakud Reservoir in the Background

urban functions. At present the commercial zone comprises of an area of about 7 hectares.

The Central Core, which is formed by the municipal and grain market is the main business focus of the town as well as of the surrounding agricultural areas. Such location is congenial for most of the residential areas. The general appearance of this business centre is not at all pleasing. The shops are located along the narrow and congested lanes which cause much difficulty to the consumers. The shops are mainly of mud structure though at some places, brick houses are used as shops.

The daily market in this zone provides the consumers of the town with perishable goods, such as vegetables, fish, meat, egg etc. which come from the surrounding rural areas. The daily commuters and farmers from the rural areas purchase their requirements of commodities while returning home. The market is devoted to both wholesale and retail trade as it is the only commercial facade of the entire agricultural district of Bargarh. A weekly market is also held here every Friday and it is the biggest in the district.

Food grains are collected from the surrounding agricultural fields by the dealers directly from the farmers and sold to the Food Corporation of India. Paddy is also collected by the rice mills from the grain market for milling.

The biggest cattle market of the State stands on the north-western part of the town, quite away from the central core. Professional cattle dealers go out to Sonepur, Patna and Kalahandi and bring herds of cattle, mostly bullocks and cows for sale at Bargarh market. Herds of buffaloes are mainly brought to Bhukta cattle market, from the border of Madhya Pradesh, about 37 km. N/NW of Bargarh and also from Chattisgarh plain (Bilaspur) . These are subsequently sold at Bargarh cattle market. The movement of cattle on the streets of the town is a regular feature. It should be mentioned in this connection that no vehicle are used for their transport even from distant places. The ribbon like extension of business establishments in the town occur along the roads and at the intersection of traffic arteries to serve the local consumers with necessary domestic goods.

The business activities of the town are controlled mostly by Marwari and Telegu people who reside in the heart of the town. The commercial establishments are favoured by ample banking facilities. There are four banks, *viz.*, State Bank of India, Sambalpur District Central Co-operative Bank, Andhra Bank and Land Mortgage Bank within the township. In general the whole appearance of the market is shabby and poor. Shops are small and bamboo-thatched. The arrangement is misleading. The roads in and around the market centre are always busy and often become congested due to mixture of slow and fast moving traffic.

Table 8.1 : Landuse Pattern of Bargarh Town, 1992

Use categories		Area in hectares
Residential	......	169.23
Administration	......	48.58
Commercial	......	8.09
Educational	......	8.09
Recreational Park and Playgrounds	......	6.07
Others water bodies roads other public uses	......	121.45
Agricultural	......	906. 88

Source: Measured by the researcher from aerial photographs

Administrative

The buildings are huddled together in a row on the central part of the town along both sides of the N.H. 6. The following offices of various government and semi-government organisations are located in the town.

1. Sub-Collector
2. Regional Marketing Officer
3. Commercial Tax Officer
4. Assistant Registrar of Co-operative Societies
5. Central Co-operative Bank (Sambalpur)
6. Assistant Director of Textiles
7. District Agricultural Officer
8. District Inspector of Schools
9. Executive Engineer (Canal Division)
10. Executive Engineer (Electrical Construction Division)
11. Telephone Exchange
12. Police Station
13. Block Development Office
14. Office of Public Works Department
15. Municipal Office
16. Sub-divisional Court
17. Offices of Private Organisations
18. Superintendent of Police Office

Panchayat Samiti Headquarter's is also located here.

Most of these offices have been constructed by the government. The semi-government and private agencies are located at hired private buildings and in many cases mixed with other types of functional uses.

Educational

A number of educational institutions have sprung up within the town to cater to the requirements of the people. They have got a distinct morphological structure though they are distributed in various parts of the town. At present there are twenty-four educational institutions of different level, of which one Degree College, 4 High Schools, 3 M.E. School, and one Secondary Training School, 8 Upper Primary Schools and 7 Lower Primary Schools are worth mentioning. Most of the institutions of lower and middle standard are located in areas of mixed uses. The high schools are located in the area adjacent to the densely inhabited localities. The Panchayat College, the only first grade degree college of the area, was established in 1960 mostly by financial contributions of the Gram Panchayats of the Subdivision. Secondary Training School is located in the out-skirts of the town. The Panchayat College is a full-fledged degree college preparing students for three years degree course in Arts, Science and Commerce. The college is managed by a Private Body. There were 612 students (including 38 girls) and 42 teachers. Both Degree College and Secondary Training, School are approachable by motor roads.

Recreational

Arrangements for recreation signify the cultural development, an healthy atmosphere of any place. But the town lags adequate recreational facilities. Parks and play grounds are few and their maintenance are poor. Clubs and libraries are also few in number and have no special characteristics. There are cinema houses and a few restaurants in the congested urban area.

Medical

Medical services of the towns are inadequate. The sub-divisional hospital, one of the oldest of the district has 25 beds, a doctor, a lady assistant surgeon, two nurses, a Dai and one midwife. It is situated in the most congested part of this town. It has no scope for further extension. Private dispensaries and clinics are few in number and are mostly confined to private buildings in form of pharmacists and druggists on the road side. One maternity and child welfare centre was set up in 1961 and a Family Planning Centre in 1958. The municipal boundary is drawn to keep provision for a vast agricultural land which would be converted into urban area in the distant future. So agriculture in the vicinity of urban periphery is an important feature of Bargarh town. To plan for its future development the authority has extended its the boundary by encroaching on agricultural land around the limit of the city boundary. As such, the area of this town subsequently has increased, covering 14 sq.km. But cultivation is still in practice on that occupied land fed by Bargarh distributary. The farmers live inside the Municipal Boundary. The tax for agricultural use of land is paid to the government and not to the Municipal Authority. They only receive tax from non-agricultural holdings.

The People

The population of the town was 22, 865 in 1971 which is about 11 per cent of the urban population of the district. Since 1911 the population is steadily increasing. (Table 8.2)

Table 8.2 : Growth of Population in Bargarh Town, 1911-91

Census	Population	Increase of population	Decadal variation of population (in P . C .)
1901	—	—	—
1911	4,684	—	—
1921	5,204	+ 515	+ 10.98
1931	7,085	+ 1881	+ 36.15
1941	8,236	+ 1151	+ 16.25
1951	9,197	+ 961	+ 11.76
1961	15,375	+ 6178	+ 67.17
1971	22, 865	+ 7490	+ 48.72
1981	35,400	+ 12535	+ 54.82
1991	51,135	+ 15735	+ 44.45

The record of decadal variation of population of this town shows some interesting features. It is quite evident from the table, that the growth is not uniform in all the decades. During 1921-31, the highest growth rate was recorded. This high growth rate was perhaps due to the inauguration of bus services in 1921 between the town and district headquarter Sambalpur. There after in 1931-41 and 1941-51 the growth rate was slow. The devastating fire in 1932 and

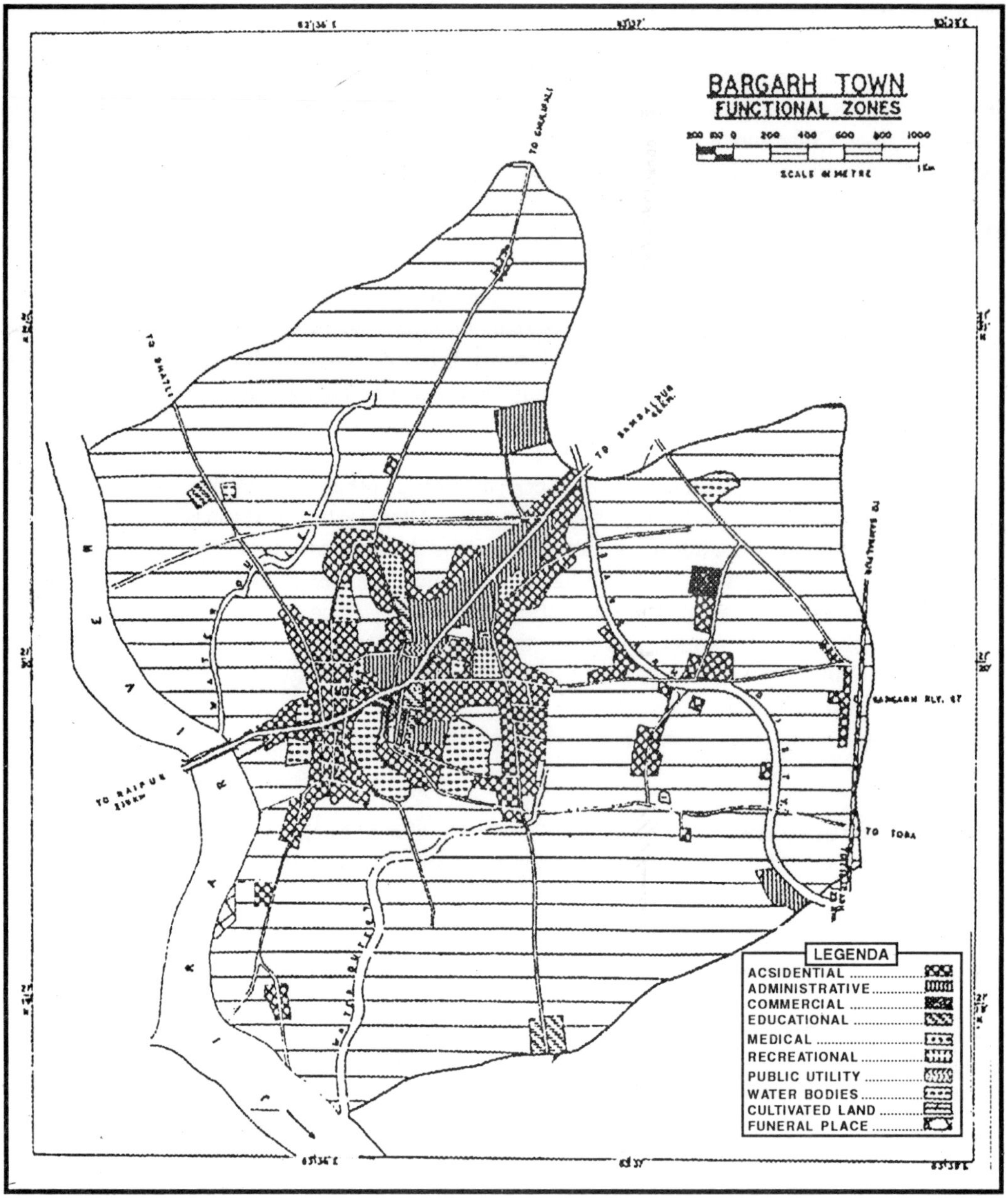

FIG. 8.5 : *Functional Zone of Bargarh Town*

high flood in the Mahanadi in 1937 might be instrumental for the low growth rate. Again high growth rate took place during last two decades. Such high growth rate may be the outcome of the establishment of Municipality, in 1951, the headquarters of Community Development in 1954, the food collecting centre of The Food Corporation of India, grain and cattle market in the town. Moreover, the successful completion of Hirakud Project and the railway service contributed much to this high growth rate.

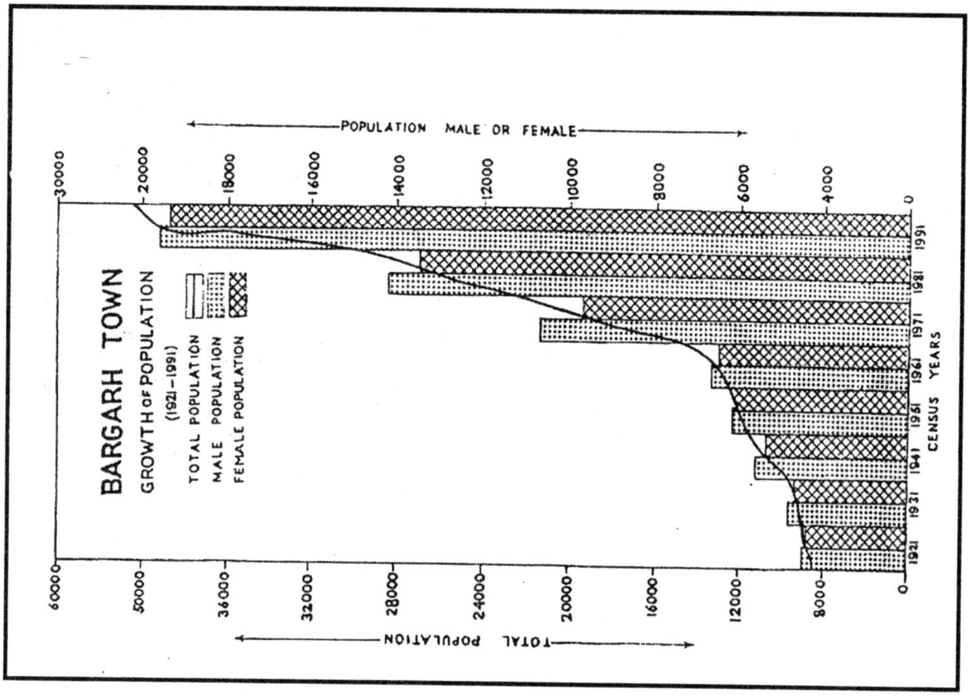

FIG. 8.7 : *Population Growth of Bargarh Town : Male and Female*

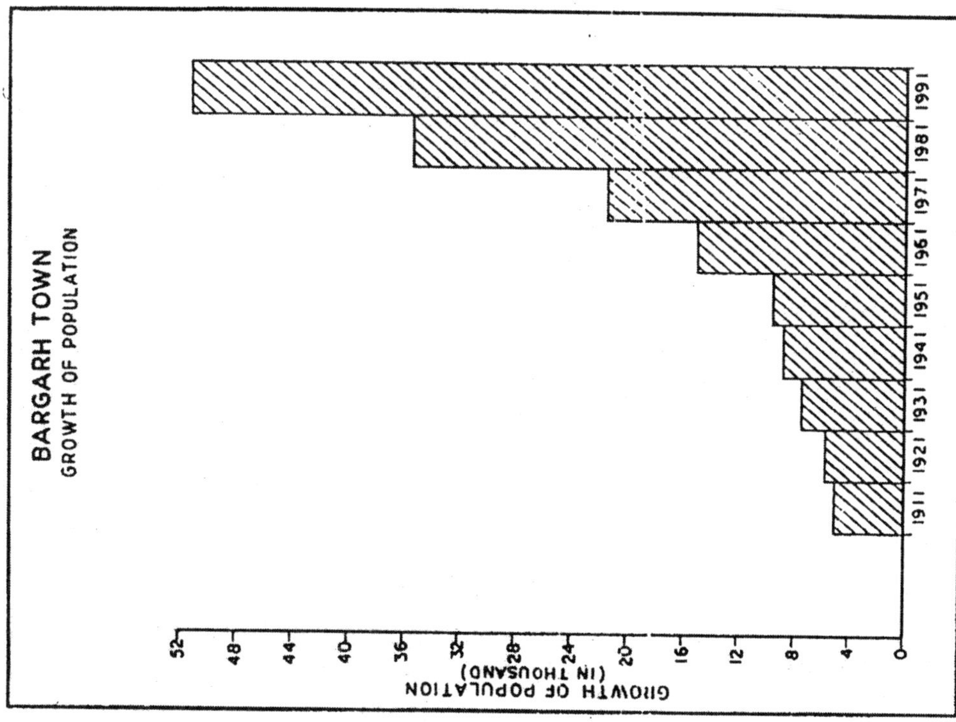

FIG. 8.6 : *Population Growth of Bargarh Town*

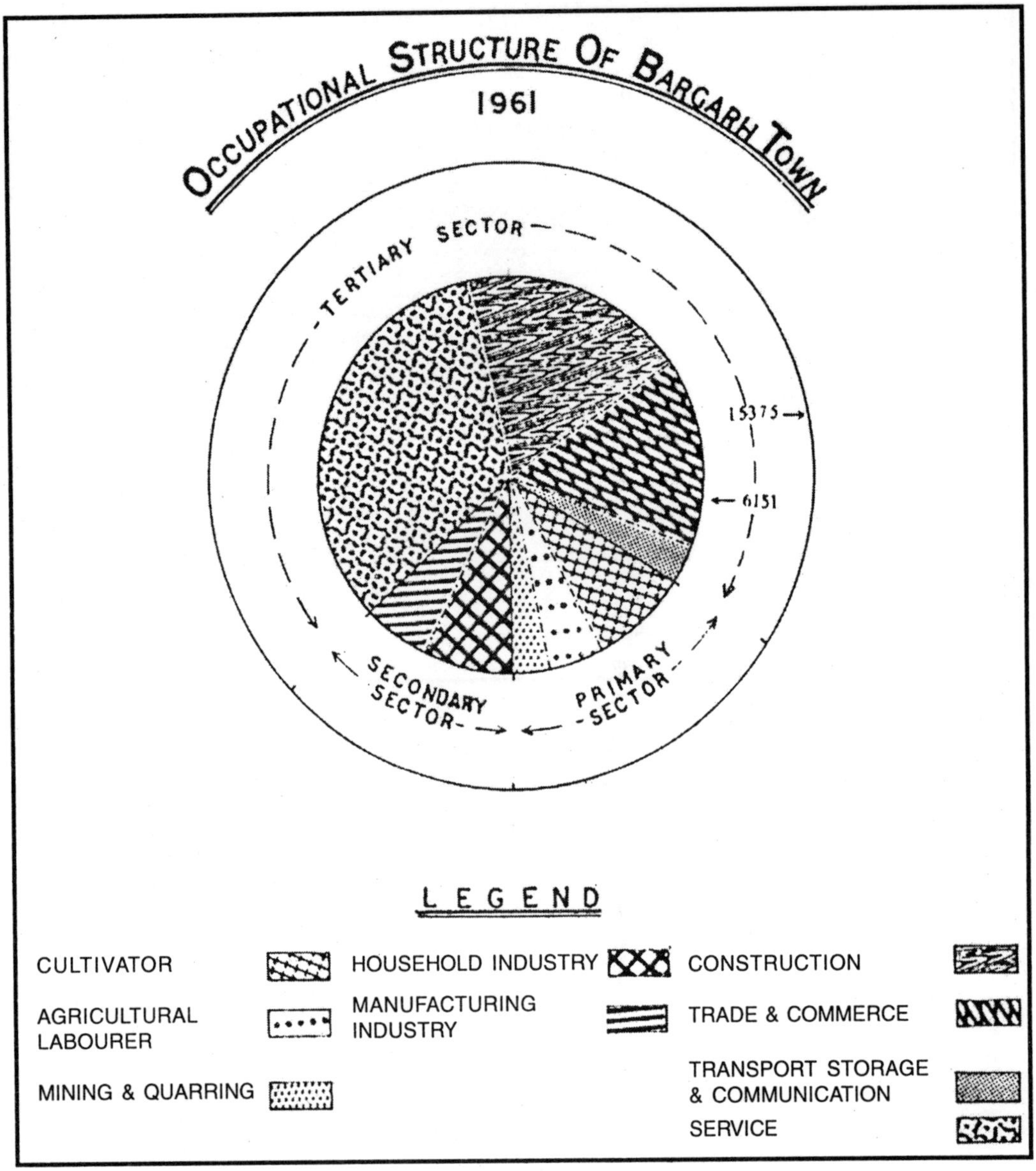

FIG. 8.8 : *Occupational Structure of Bargarh Town, 1961*

As regards density, the town records interesting trend. During the first half of the century, density increases considerably. In the next ten years that is during 1951-61 the area of the town was increased from 5.96 sq.km. to 13.21 sq.km. and the density of population shows a decreasing figure than the previous decades. But the overall population inereased from 15, 375 in 1961 to 22, 865 in 1971 and density of population in 1971 was recorded as 1,731 per sq.km. (Table 8.3).

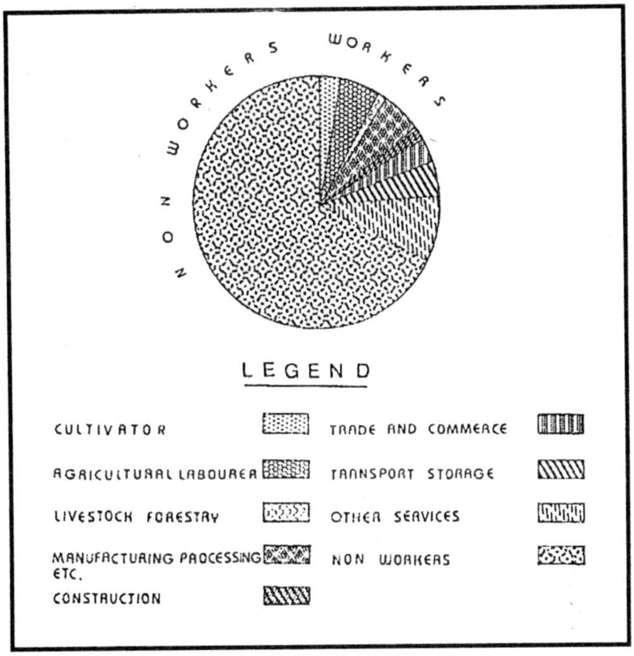

FIG. 8.9 : *Occupational Structure of Bargarh Town, 1971*

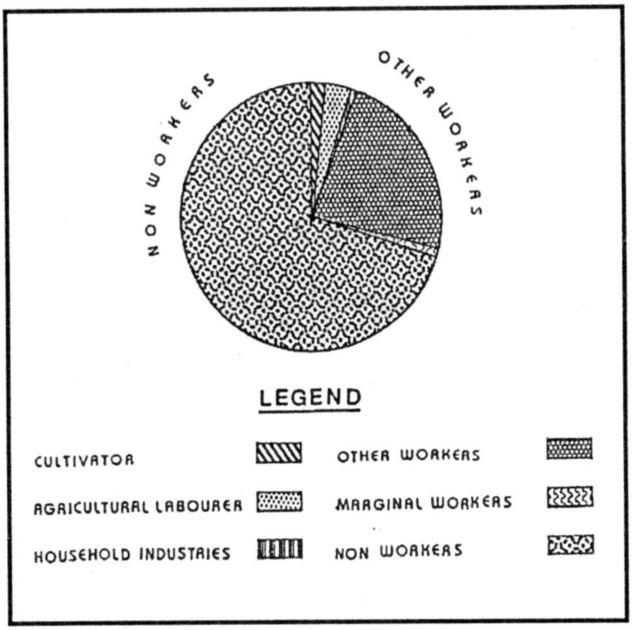

FIG. 8.10 : *Occupational Structure of Bargarh Town, 1981*

Table 8.3 : Density of Population of Bargarh Town, 1911-91

Census years	Area in sq.km.	Population	Density/persons sq.km.
1911	5.96	4684	786
1921	5.96	5204	873
1931	5.96	7085	1189
1941	5.96	8236	1382
1951	5.96	9197	1543
1961	13.21	15375	1164
1971	13.21	22865	1731
1981	13.21	35400	2680
1991	19.21	51135	2662

The town is divided into 15 wards. The density figures of these Wards are furnished below based on 1991 population approximately (Table 8.4)

Table 8.4 : Density of Population of Bargarh Town by Wards

Ward No.	Area in hectare	Population	Density/hectare
1.	17	2634	155
2.	17	2820	166
3.	25	2160	86.00
4.	7	905	129
5.	5	2167	433.6
6.	12	1683	140.0
7.	110	2272	21
8.	15	4088	273
9.	25	3927	117
10.	56	6702	120
11.	227	6661	29.0
12.	202	5045	25
13.	72	4819	67.0
14.	186	1717	9
15.	345	3533	10

Ward Nos. 2, 5 6 and 8 record the highest densities and are the most congested parts of the town. Density of these three wards are 166, 433 & 273 per hectare respectively. The population concentration in Ward Nos.1, 4, 6 & 10 is somewhat moderate. The rest of the Wards are very thinly populated particularly the Ward Nos. 14 & 15 as most of these wards are still agricultural.

The town has experienced a rythmic flow of people from different parts of the country since the early stage of its development. The most interesting demographic feature of the town is that Marwari, Sindi and Telegu peoplo constitute a large percentage of the population while the town is located well inside Orissa.

Functional Characters

The prevalence of non-agricultural pursuits among the population is the main criterion of urbanisation and the activities which provide the earning of the people are the good indicators of urban functions. The nine-fold industrial classification of workers of the town in 1961 and 1971 are arranged into three broad categories in order to highlight the basic changes which have taken place during the span of 10 years. These are (1) Primary Sector (2) Secondary Sector, and (3) Tertiary Sector of employment.

PLATE 5 : A Distant View of Hira Cement Factory in Bardol near Bargarh Town.

PLATE 6 : An overview of Bargarh Canal near Burla

Primary Sector

This includes people engaged in cultivation, agricultural labour, live-stock, forestry, fishing, hunting and plantation, mining, orchard growing and allied activities. In 1971, about one-fifth of the total working population were engaged in this sector, as against 16 per cent in 1961. In the primary sector agriculture dominates over livestock, fishing, forestry and mining activities – as agriculture is still practiced within the municipal boundary. It is interesting to note that employment in mining activities has steadily decreased in the last decade. In 1971, mining provided employment to only 3 persons as against 187 persons in 1961.

Secondary Sector

This includes both household and manufacturing activities of the town. In 1971, about 14 per cent of the working population were engaged in different industrial activities of the town as against 13.32 per cent in 1961. Pottery, weaving and mat making are the major household industries in this town. Manufacturing industries include cement, sugar and rice mills, wooden furniture making and bidi making.

Employment in manufacturing industries has considerably increased, *viz.*, 541 persons in 1971 against 348 in 1961. The town has enough prospects for industrial growth in near future.

Tertiary Sector

This sector comprises of trade and commerce, construction, transport and services. This sector provides employment to about 70 per cent of the total working population in 1971.

Trade and Commerce

This function plays an important role in shaping the economic growth of the town. It may be considered as the predominant function of this town.

Construction

The employment in this function has steadily declined from 1175 persons in 1961 to only 233 persons in 1971. The figure would suppose to rise again following the development of this town.

Table 8.5 : A Comparative Study of the Occupational Structure of Bargarh 1961-81

	1961			1981		
Total population		15,375			22,865	
Workers	6151	or	40.06%	6898	or	30.70%
Non-Workers	9924	or	59.94%	15967	or	69.30%
1. Primary Sector	1970	or	15.77%	1481	or	21.48%
2. Secondary Sector	819	or	13.32%	941	or	13.64%
3. Tertiary Sector	4362	or	70.91%	4475	or	64.88%
1. (a) Cultivators and Agricultural	783	or	12.75%	1280	or	18.56%
1. (b) Livestock, Fishing, etc.	Not classified in 1961 census			198	or	2.78%
1. (c) Mining and Quarying	187	or	3.046	3	or	0.04%
2. (a) Household Industries	471	or	7.66%	398	or	5.74%

Contd.

Contd.

	1			2		
2. (b) Other than household industry	348	or	5.66%	545	or	7.90%
3. (a) Construction	1175	or	19.10%	233	or	3.38%
3. (b) Trade and Commerce	940	or	15.28%	1811	or	26.25%
3. (c) Transport Storage and Communication	195	or	3.17%	778	or	11.28%
3. (d) Other services	205	or	33.63%	1654	or	23.98%

Source : Computed from the records of the Census of India, Orissa for 1961 and 1971

Infrastructural Facilities

The municipality installed one 10 H.P. water pump to provide pipe-water for the town. Water supply was quite inadequate and was stopped. At present the public health department supply pipe-water to the town by installing more powerful pumps. The municipality maintains 27 wells. There are 19 tanks within the municipal area.

This local body has provided 870 m. of *pucca* drains and 28 km. of roads. The municipality has provided electric light, throughout the town and kerosene street lamps in the small streets. It maintains also a children's park. The town has an Inspection Bunglow managed by the Public Works Department and a well furnished Rest House. There is, however, a Dharmasala for the common people. The town is provided with Post office, Telephone Exchanges with 100 lines and Police wireless station.

Communication

The town serves as a nodal point of various road networks. The NH 6, the main Trunk Route between Calcutta and Bombay passes through this town. The town is also connected with Cuttack and Bhubaneswar by NH 42. which meets NH 6 near Sambalpur. The State Highway No. 2 or Bargarh-Bolangir-Bongumma Road takes off from the NH 6 just beyond the Jira Bridge at Bargarh and moves towards Bolangir. Some of its feeder roads like Sohela-Barpali-Binka Road, Bejapur-Charmunda Road and Kotapali-Barpali-Kumbhari Road connects the town with other important places in the south-central part of the district. The major District Road No. 33 or Bargarh-Bhatli-Ambahona Road starts from NH No. 6 at Bargarh and hence connects the town with important places of the north central part of district.

Moreover, within the town area there are roads of various types which are mentioned by Municipal Authority.

Table 8.6 : Types of Road of Bargarh Town

Types of Road	Length in km.
Black topped	5.3
Cement conercte	0.30
Metalled	11.57
Earthen	5.09
Other Roads	7.11
Total length	29.37

PLATE 7 : Bargarh Town, Market Centre

PLATE 8 : Inside view of Market Centre of Bargarh Town

Vehicles and Conveyance

Motor vehicles, cycles and cycle-rickshaws are the principal mode of conveyance to the passengers of the town. For goods flow from one part of the town to another mainly, bullock-carts are widely used. Sometimes motor vehicles also ply for the transportation of goods in the town.

Table 8.7 : Types of Conveyance at Bargarh Town

Year	Cycle	Cycle-rickshaw	Bullock-cart
1962-63	303	59	34
1987-88	669	73	12

Most routes converge to the town area and area under the Central and the State Transport Department although some private concerns are operating. Out of three units of the State Transport Department, Bargarh is one and falls under the Bolangir Zone.

Railway Network

A comparatively new rail link passes through Sambalpur and Banargarh Subdivision. It was initiated on April 15, 1963. Bargarh is thus connected by rail-nets with Jharsuguda, Lapanga, Rangali, Sason, Sambalpur Road, Sambalpur, Hirakud and Barpali.

A new metre gauge railway line has been opened between Bargarh Dungri on the 14th April, 1967. The line is 53 km. long and has been built by Industrial Development Corporation of Orissa mainly to carry 2000 tonnes of limestone daily from Dungri Quarry to the Cement Plant site at Bardol only 3 km. off Bargarh. There are four stations on the line, *viz.*, Gopalpur, Somardhara, Kusumda and Dungri.

Future Development of Industries

Though the town itself is not an industrial one, its neighbouring areas are being heavily industrialised. Only 3 km. to the north and adjacent to Khaliapali village Hira Cement Factory is located, which went into production since 1968. About 5 km. to the east near Tora village there is a spining mill. A sugar factory has been recently started. Industries like rice milling, weaving, bidi making, carpentry, pottery, black smithy, metal works are there in the town.

Industrial potentiality of the town has radically changed after the construction of the Hirakud Multipurpose Project. Hydroelectric power is available from Hirakud. These factors have influenced both agrobased and other metallurgical industries of the town. Forest and clay reservoirs in the region are potential raw materials for forest based and ceramic industry in the coming years.

There is also a project for starting ancillary industries at Bargarh to feed requirements, already established in and around the town.

Market-Town

Bargarh is a Subdivisional headquarter. This market is connected by all weather *pucca* road with Sambalpur, the district headquarters, and by railway with Jharsuguda Junction on Howrah-Nagpur line. Bargarh market is also connected with the producing areas of Sonepur and Bolangir subdivision of Bolangir district by all weather communications. It has also trade activities with Saraipali, Basria and other contiguous areas of Madhya Pradesh, because of easier communication available to those markets. On account of convenient situation, Bargarh market has the capacity to draw appreciable quantities of agricultural commodities from different markets of this district and also from outside markets. This market serves the entire Bargarh Subdivision besides parts of Bolangir District and some villages at Madhya Pradesh.

Due to its location in the midst of prosperous farm lands, Bargarh has now become a centre of Regional Marketing Co-operative Society. The cotton handloom products of the region is important. Die-processing on the textiles of the region is very famous which attracts lots of foreign tourists who give the local weaver community advance booking for availability of the best product from the economically weaker sections of the weaver community. Recently a fairly big rice-cum-oil mill has been set up. At present, the activities of the Society covers 10 Panchayat Samities consisting of more than 60 Grain Golas (storage bins). The marketing co-operative society of Bargarh is mainly engaged in the distribution of chemical fertilizers to Grain Gola Co-operative Societies, who in their turn distribute the same to the producers of their areas. The Society also undertakes procurement of rice and paddy under State Trading Scheme and also function as authorised dealers in cement, wheat and other consumer goods etc.

The advent of Hirakud Irrigation Project has brought about a new era in the agriculture of Orissa as well as in the economy of our country. Enormous development in the agriculture has been made and the whole command area has led to the formation of three agricultural districts of which Bargarh ranks first with the title "Granary of Orissa." The Agricultural District of Bargarh consists of 11 Blocks with about 3, 000,000 hectares of arable land. Bargarh is acting as a nerve centre by providing all sorts of help. There are two basic development objectives, *viz.,* to create sound economic development of the region and to develop inter-related economic structure by establishing close relationship between the farming people and the market town. The road linkages between the town with the adjacent lands thus, assure the farmers essential infra-structural facilities like credit, vocational training, agricultural extension work and medical facilities. The headquarter of Community Development is located in the town. The farming people are benefitted by farming implements, manure, informed technical advice and various other forms of co-operation for improvement of agricultural yield.

BURLA TOWN

Location

Burla (Lat. 21° 30' N and Long. 83° 50' E) is a small town on the right bank of the Hirakud Project in Sambalpur District of Orissa. The nature of its development has received the attention of sociologists, economists, geographers, etc. The present township is formed on the areas of 2 villages of Burla and Kirba. In 1951, the population of these villages were 3,791 and 528 respectively. With the construction of Hirakud Dam Project in 1947, their rural character has been transformed to a rapidly growing service town of the project area. The original villages are situated just on the right bank of the Mahanadi river. The said villages have also been shifted to about 1.6 sq.km. south of its old position due to construction of this dam.

The area over which the township has been developed is bounded by the Mahanadi river on the north and east, Chandli Dungri forest in the west and the rural landscape in the south. In 1961, it came under Notified Area Council (NAC). The NAC boundary covers Burla, Kirba, Budamunda and Gulgunda villages. But Budamunda village has now been submerged by the power channel of the project. According to census classification Burla has been a Class IV town in 1971 with an area of 17.6 2 sq.km. Rapid development of the region has now made this small town a beautiful urban centre with a promising future. Prior to the construction of Hirakud Dam, the Mahanadi river was notorious for floods every year. But the floods have now been checked. The area is now getting canal water facilities for farming. Burla has come to prominence during the phase of construction of Hirakud Dam. The selection of the area for the growth of the town is due to flat terrain and accessibility to district headquarters and other important places by NH 6.

Physiographic Setting

The township is situated on a flat plain in the north-west of Sambalpur district. It is part of the flood plain of the Mahanadi river. The average height of the town is 150 metres above the sea level. The general landscape is beautiful, as groves and orchards are found in the fringe of the town. There are few isolated low hills and rocky outcrops in and around the town. The highest point near Burla town is 310 m. in the northwestern part. It is a long ridge covered with forests named Chandli Dungri Reserve Forest. At the foot-hills, afforestation is going on. But some portion of the forest has now been cleared for the lateral expansion of the town. In the southern part of the town, the new buildings of the Sambalpur University are coming up. The Jawahar Minar is on the N. W. part of Chandli Dungri forest. It faces the Hirakud Reservoir, thereby attracting tourists from all over the country. There is also a small isolated hill in the Southeastern part of the town. In the northern part of the town lies Mahanadi river which is exposed below the Dam as dry rocky beds.

The area enjoys monsoon rain. In general the temperature is not excessive but the heat is aggravated during the summer months by radiation from the sandy and rocky beds of the Mahanadi. The climate is characterised by hot dry summer, though the summer temperature is lowered by 2°-3°C due to the influence of Hirakud reservoir. The rain is generally well distributed during the south-west monsoon season. May is the hottest month (41. 6°C) and December is the coldest (20.9°C) . The average rainfall is 1250 mm. The monsoon is generally regular, sometimes it may delayed by 2/3 weeks. Such discrepancy is normal for the monsoon of South Asia. During the rainy season a number of depressions from the Bay of Bengal cross the east coast and move in north-westerly direction, due to which this part gets heavy rainfall and gusty winds.

Origin and Growth Pattern

Burla is originated as the outcome of Hirakud Dam. This industrial set up is associated with favourable infrastructural facilities, development of hydroelectric generation of the power plant constructed by the site of the Dam and nearness to the raw materials. With the inauguration of Hirakud Project, the economy of the region has undergone rapid transformation. Irrigation has brought a new era of the activity for its development. Primary activities can not be said to be main factor for the growth of this town. The town itself is flourished to facilitate, civic amenities. Before 1951, there was no town on its present location. Only a few office buildings were there to look after the construction work of the Dam. With the initiation of Hirakud Project the Burla township has been set up for smooth functioning of the project. Since then the town is passing through successive stages of development.

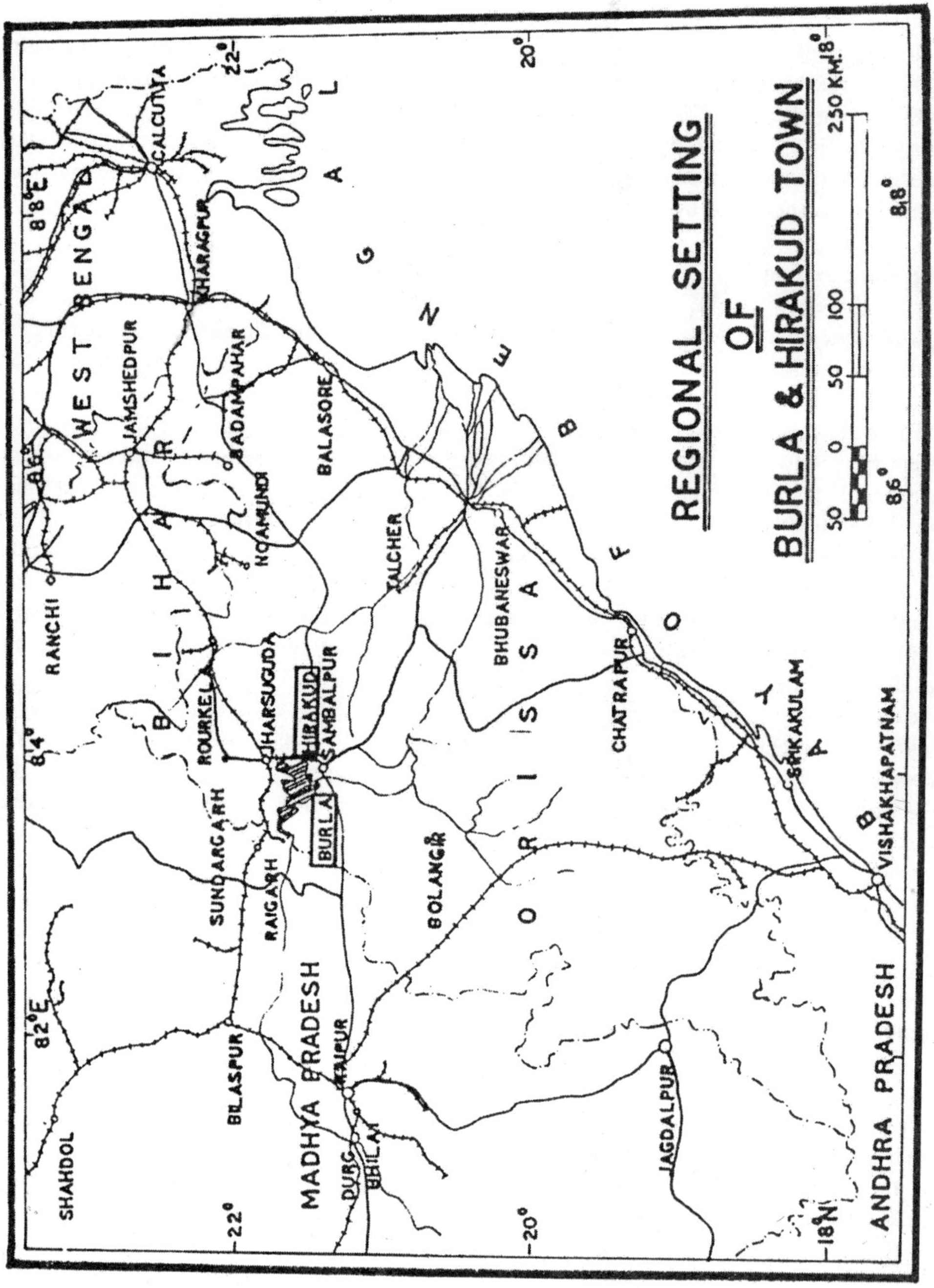

FIG. 8.11 : Location Map of Burla and Hirakud Towns

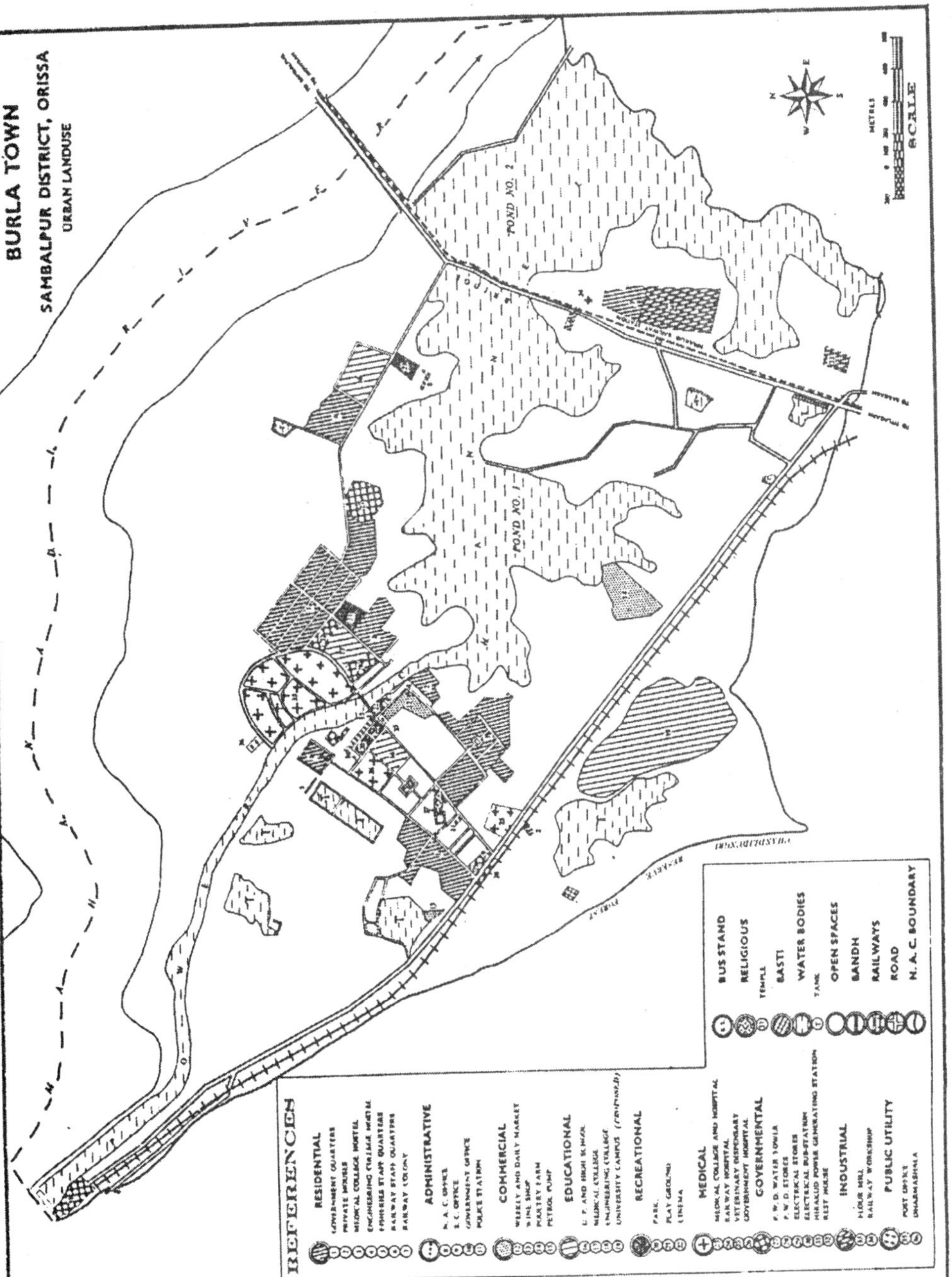

FIG. 8.12 : *Morphology of Burla Town*

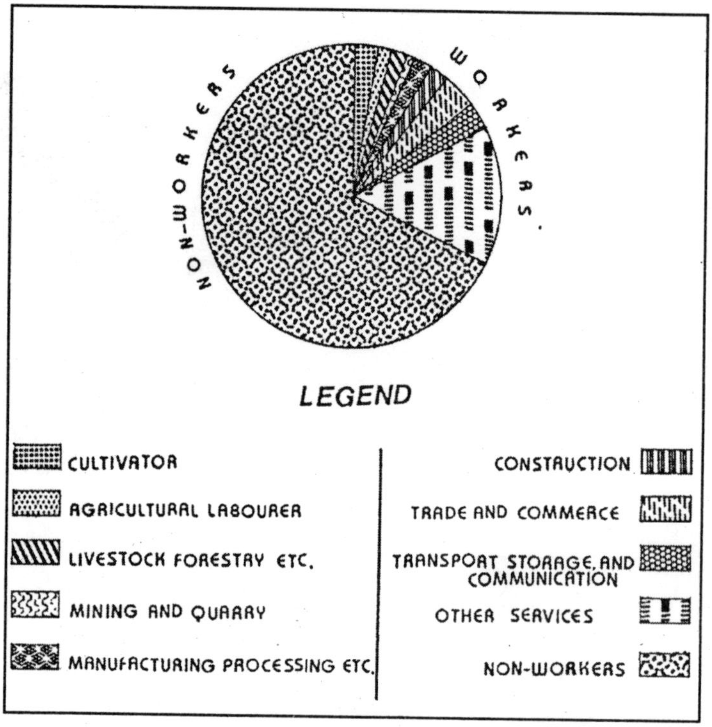

FIG. 8.13 : *Occupational Structure of Burla Town, 1971*

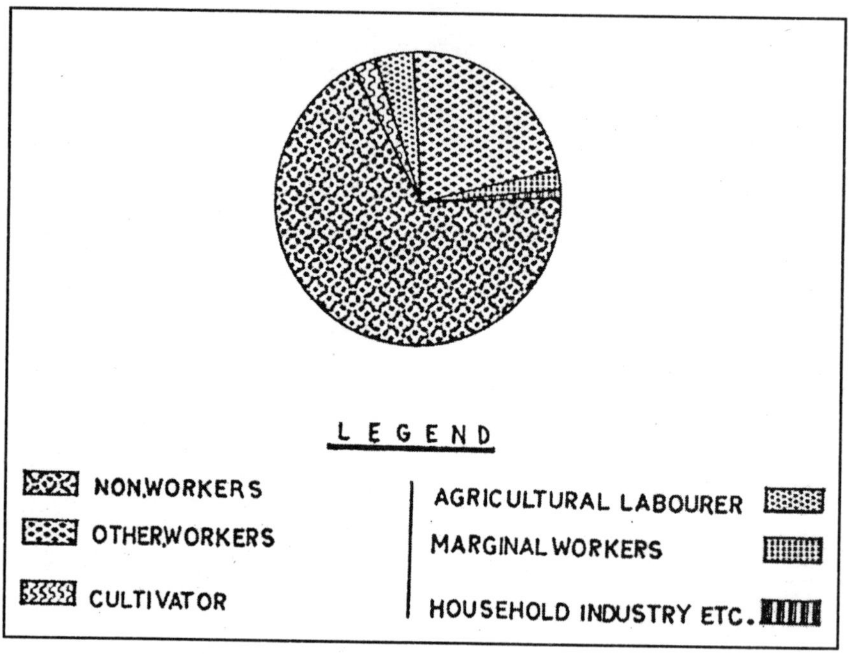

FIG. 8.14 : *Occupational Structure of Burla Town, 1981*

PLATE 9 : Main Gate of Sambalpur University of Burla Town

PLATE 10 : Earth Science Department of Sambalpur University

In 1951, the total population of the town was only 4,319 (Burla 3,791) and (Kirba 528) . The enormous movement of population into the area for the construction of dam helped in the growth of the township. Due to rapid increase of population by fresh immigration from the neighbouring areas, the total number in 1961 stood at 10,230 which caused a net increase of 5,911 persons in 10 years or 591 per year. The increase rate of population during 1951-61 was 46.61 per cent. In 1971, the total population of Burla township had swelled to 15,587 i.e., a total increase of 5,357 persons during 1961-71. The rate of population growth during 1961-71 was 53:57 per cent or 5.36 per cent per year. The growth rate increased with the development of this township.

Table 8.8 : Decadal Variation of Population of Burla, 1951-91

Year	Total population	Decade variation	Density per sq.km.	Growth rate (in per cent)
1951	4,319	—	245.1	—
1961	10,230	+ 3911	580.5	46.61
1971	15,587	+ 5357	884.6	53.57
1981	27,882	+ 2295	1582.4	78.0
1991	34, 635	+ 6753	1965. 6	24. 0

The poor villagers had to migrate to the outskirts of the town areas, though some of them preferred to stay back. They are now engaged in trade and commerce. Low paid workers of the area are unable to maintain the high standard of living of the urbanised area.

In 1951 the population density was 245.1 persons per sq.km. and with the rapid growth of the town the density accordingly increased to 581 persons per sq.km. in 1961. The corresponding. figures for 1971, 1981 and 1991 are 885, 1582 and 1966.

Table 8.9 : Density of Population of Different Wards of Burla, 1981.

Ward No.	No. of occupied residential houses	No. of total persons	Percentage of total population of town in the ward	Percentage of educated persons to the population of the ward
I.	188	2800	10.0	24.48
II.	142	1050	3.8	23.68
III.	193	1404	5.0	57.45
IV.	442	2763	10.0	71.73
V.	407	3316	12.0	74.27
VI.	209	2208	8.0	57.26
VII.	517	3404	12.0	26.21
VIII.	148	808	3. 0	34.17
IX.	503	1617	0.2	0.33
X.	162	1736	6.2	25.3
XI.	219	1769	6.3	35.3
XII.	95	1819	6.5	56.36

Source : Compiled by the Researcher, from the records of Census of India

PLATE 11 : Burla Medical College

PLATE 12 : Overview of Hirakud Town

Maximum concentration of population occurs in Ward No. VII, being followed by Ward Nos. I, IV, V, VI and XII. Ward No. IX shows a marked population deficit area of the town occupying only 0.33 of the total population. The town is not congested and density is moderate only 1966 per sq.km. in 1991.

According to 1991 Census, the town has the sex ratio of 850 females per 1000 males. The total number of Scheduled Caste was 4465 and the Scheduled Tribe was 2847 persons. The town has got modern educational facilities having the most modern Medical and Engineering Colleges of the State. Of the total population of about 34,635 in 1991about 55 per cent persons are literate.

Table 8.10 : Occupational Character of Working Population of Burla Town, 1961-1981

		.1961	1971	1981
Total Workers		4,587	4,914	8,089
I .	Cultivators (in %)	8.72	7.73	8.25
II.	Agricultural Labourer to total workers (%)	1.25	3.4	12.9
III.	Mining and Quarrying to total workers (%)	2.65	2.00	
IV.	(a) Household Industry (% to total workers)	4.69	7.00	1.00
V.	(b) Other than Household Industry (in %)	5.58	5.00	
VI.	Construction (in %)	5.88	6.30	
VII.	Trade and Commerce (in %)	5.95	11.00	
VIII.	Transport and Communication (in %)	3.24	9.00	
IX.	Other Schemes (in %)	62.11	48.00	
X.	Other Workers	Introduced in 1981 Census	Introduced in 1981 Census	74.00 III + IV V(b), VI to IX
XI. Marginal Workers		-do-	-do-	6.00

Source : Computed from the Records of the Census of India.

Table 8.11 : Demographic Character of Burla Town

	1951	1961	1971	1981	1991
Total population	4,319	10,230	15,587	27,882	34,635
Percentage of male to total population		59.95	58.0	54.6	54.0
Percentage of female to the population		40.15	42.0	45.4	46.0
Percentage of S.C. to total population		17.09	16.5	19.14	12.8
Percentage of S.T. to total population		1.68	7.5	9.9	8.3
Percentage of literate to total population		43.56	53.73	55.2	45.5
Percentage of workers to total population		44.88	31.6	29.0	24.0

Source : Compiled from Census of India

Occupational character of this town is very interesting. Of the total population in 1961, 44.88 per cent was workers and 55.12 per cent non-workers. Among the total workers, 4, 587 people were engaged in different jobs. Female workers were negligible in number only 792 of the 4, 587 workers in 1961.

Functional Zones

As per occupational structure it shows that Burla is a service town, though a sizeable portion of her population are engaged in education as students due to establishment of the Burla Medical College and Engineering College. The educational character is further strengthened after shifting of the University of Sambalpur here. The building of the University is under construction.

The so called civic administration is controlled by the Notified Area Council, though the Council is still in a rudimentary condition, as the town itself has come into existence in recent period, after the implementation of the Hirakud Dam Project. Most of the State Government offices belong to those of Irrigation and Power. These offices are the pivot of the town controlling the whole system of the Hirakud Project regarding maintenance of the Dam and distribution of power and water. The power house is just situated by the side of the Dam of the left bank of the Mahanadi with the total installed capacity of 198 MW. Commercial centres also play some important role in the economy of the area. Both retailers and wholesellers at present are doing good business. The prices are generally high in comparison to the normal prices in Sambalpur town. The functional zones of the town can be classified as below :

The residential landuse has been planned in a systematic manner. For the convenience of administration, the town is sub-divided into 9 wards. Different organs of the township as schools, colleges, institutions, parks, offices, hospitals are found in different parts of the residential areas. Some of the quarters are arranged in a rectangular fashion along both sides of the street. In 1961 the total number of occupied residential house was 2, 604. The residential zone includes Government Staff Quarters, fisheries staff quarters, and some private quarters. Since then, the number of residential houses has been increased to 3,138 in 1971. Railway quarters are situated near the Hirakud Railway Station.

The main function of the town is administrative for the maintenance of Hirakud Dam Project which covers substantial part of the town. The administrative units also include N.A.C. Office, S.C. Office, other government offices and police station. These administrative offices is situated in the southern part of the town for the maintenance of law and order. The N.A.C. has the supreme power to control all the urban functions.

The commercial landuse is controlled by the administrative body of the town. To curb undesirable congestion and maintain cleanliness, the haphazard growth and development has been completely restricted. The market centre of the town is located almost in the central part of the residential zone to make it convenient. For the customers from all concern, there are some permanent shops which serve the essential commodities of the buyers. Book stalls, stationary, tailoring, textile, laundry, restaurant, hotel etc. are common features of the market. During the construction period some shop workers served in the project. Some vegetables come to the market and are of local nature. Fishes are supplied from the Hirakud reservoir, but most of the fishes are transported from here to Sambalpur town. The poultry products come from the poultry situated in the southern part of the town.

The weekly market occurs in the same site of the daily market, people come with different commodities from remote areas. People from the periphery of the town also come to buy necessary commodities. In this weekly market huge quantities of rice and lesser amount of pulses are sold. It is also interesting to note that different domestic animals like cattle, goat, etc. are also sold here. As a whole, the weekly market is of greater importance to the local people of the township and its outskirts than daily market. The railway colony which is situated near the railway station has separate miniature market.

The centralised institutions of the town are symbols of high urban rank. These institutions are maintained by the local authority as well as by the Orissa Government. The institutions are

PLATE 13 : Aqueduct to check the flow of water of Bargarh Canal.

PLATE 14 : Viaduct to Carry Water through Concrete Channel over Natural Stream

co-ordinated with plans for better communication --- park and playground. The institutions are Upper Primary, High Schools, Medical College and the University Campus. The Primary and High Schools are situated at the heart of the residential area. There is a park facility for the school students. Metalled road surrounds the school. The medical college lies in the northern part of the hospital . It is also well connected by metalled roads with all parts of the town. It is the second Medical College in Orissa and was established in 1959. The government is responsible for the management of the Medical College. The first Engineering College of Orissa has been set up at Burla in 1955. It lies at a distance of about 2 km. from the planned residential area and is connected by only one metalled road with the heart of the town. The Sambalpur University Campus is being built in the southern part of the town demolishing part of a small hillock.

Parks, playgrounds, cinemas, open spaces of various kinds are considered as recreational units. The Jawahar Minar is situated just outside the north western, part of the N.A.C. boundary near the power station facing the reservoir. The State Government is maintaining a small luxurious hotel, (Ashoka Niwas) which attracts a large number of tourists. Post office, Dharamsala, Rest Houses, etc. are the units of public utility category. Medical College Hospital, Railway Hospital and Voluntary Dispensary are the important units of the group medical centre. The Medical College Hospital is situated between the Hirakud reservoir and the Mahanadi River in the northern part of the town. It has been assumed that the site has been chosen for the better health and sanitary conditions of the patients. Nurse, auxillary nurse and *Dai* training centres are also included in the Hospital management. The number of beds is only 38 in 1961. The area of the Medical College Hospital is very big and there are provisions also for future development. The railway hospital in the colony serves the railway staff. There is only one temple located in the town and a site is marked for a mosque.

As manufacturing is not the principal function of the town, a small patch of land is kept for this purpose. Flour mills, railway workshop, Hirakud Reservoir Power Generation Station, etc. are located within this patch of land. The Right Bank Power House did not exist in 1951 but came into existence now.

A number of tanks locally known as '*Kota*' built earlier for the purpose of local irrigation are found in different places of the town. The configuration and the sub-soil condition of the area help for the construction of the tanks. The main bodies of the town are the two big tanks connected with the power channel which has been built to divert water of the reservoir through this channel after generating electricity.

Communication

Burla Town is well connected by metalled roads with other parts of the State. The station is 5 km. from the town. A small line is there along the southern margin of the town which runs up to the power House though not functioning at present. Running parallel to the railway line is NH 6 which connects the town with Sambalpur and Raipur. Inside the town, there is no intra-city bus service. Taxies ply regularly to Sambalpur, Bargarh, Bolangir etc.

It is worthwhile to mention that though the town possesses unique facilities regarding education, medical treatment, etc., the people cannot enjoy the urban advantages properly due to lack of co-ordination among the different societies particularly between the student communities and service holders. Business professionals are very few. The prices of consumers goods are much higher than in Sambalpur. Many well-to-do families visit Sambalpur for marketing. It is anticipated that with the growth of population, small scale industries may develop here and these may be associated with the Industrial Estate situated near Sambalpur by the side of the NH 6.

PLATE 15 : Hirakud Reservoir showing Conical Hills on the other side.

PLATE 16 : Concrete Dam of the Hirakud Reservoir

HIRAKUD TOWN

Hirakud, a small town having an area of 8.81 sq.km. has been formed at the Hirakud Project site on the left bank of the Mahanadi River. It is well connected by motorable road with railway line. There is also a railway yard for loading and unloading of raw materials and finished products. Hirakud Railway Station lines are situated at a distance of 6 km. from the main town. The area is served by Rourkela-Titlagarh railway line under South-Eastern Railway line. Calcutta-Bombay NH 6 is only 6 km. from Hirakud township. The Administrative Officer's Training School, a Poly-Technique Institute, Aluminium Factory and Cable Factory are the most important organs of the town. The town has a Notified Area Council since 1962. In 1966 the Town Planning Office under the Urban Development of Orissa Government prepared a Master Plan of Hirakud covering an area of about 604 ha. The town in 1961 was in Class V category. With the rapid urban growth, it changes its rank to Class IV category in 1971, with a population of 15, 040.

Location

Hirakud (Lat 21° 30' N; Long 83° 50' E) is a small town in Sambalpur District of Orissa and the township developed at the project site. Formerly it was a small island village of Mahanadi, but now the new township has been shifted on the east bank. On the island is a beautiful garden maintained by the Hirakud Dam Circle. Original Jamda Village became Hirakud town. The village was well populated even in 1951 consisting of 6, 559 persons, due to construction of Dam. Jamda village became a busy centre of activity during the construction period and the landless people in greater numbers settled down in and around the village in search of employment. The town is bounded in the north by Hirakud Reservoir, dry bed of Mahanadi in the south and west and rural small village in the east.

Physical Setting

The town has developed on a flat area with an average height of 151 metre from the sea level. It is also within the reach of the Mahanadi flood plain. A little hillock in the north stands abruptly where Gadhi Minar is erected with a beautiful garden. The shallow soils are mostly red and yellow. The climate is the same as Burla. It lies on the track of cyclonic storm from Bay of Bengal. In short the town is surrounded by highlands into north and east and there is a hill in the south covered by Lachmi Dungri Reserve Forest. Between the Lachmi Dungri and northern high land there is a gap which is itself a continuation of the flat area of the town to the eastern plain.

Origin and Growth Pattern

The economy of the area has been changed considerably after the construction of the Dam. The power and water supply of the project have helped in the growth of the town. In the past Hirakud was an island between the two branches of the Mahanadi 10 km. north of Sambalpur. The place is so named as diamond (Hindi-Hira) quarrying is said to have carried out in the past in the river bed. But the actual township has now developed at the project site and the original island is uninhabited. Original Jamda village gradually grew at the site of the existing Hirakud town. Since 1961 it has got the status of a town in the category No. VI with a population of 8,593 with an area of about 9 sq.km. With the successive stages of development, it became a Class V town with a population of 15, 040 in 1971. Thus, the rapid growth of this town attracts the observers. The town has a Notified Area Council since 1962. The importance of the town has gradually increased with the development of the Hirakud Project as large number of industrial establishments particularly aluminium industry have been set up here.

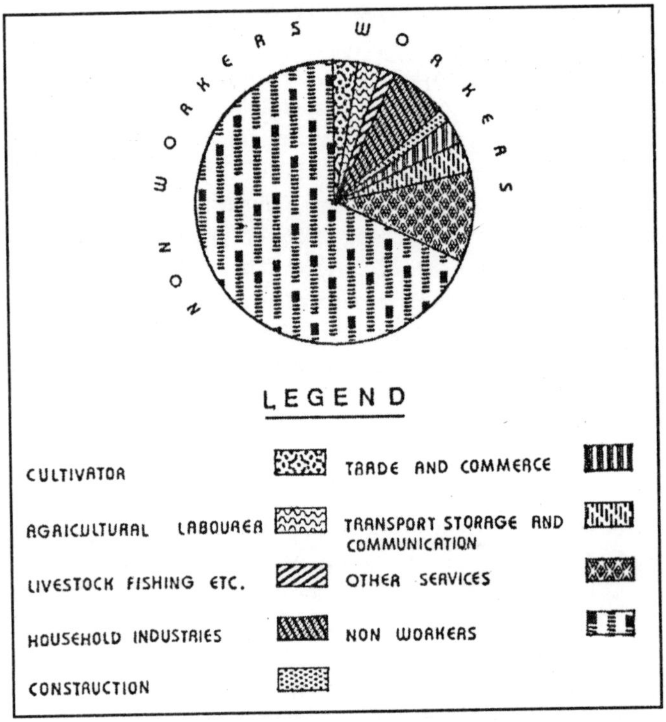

FIG. 8.15 : *Occupational Structure of Hirakud Town, 1971*

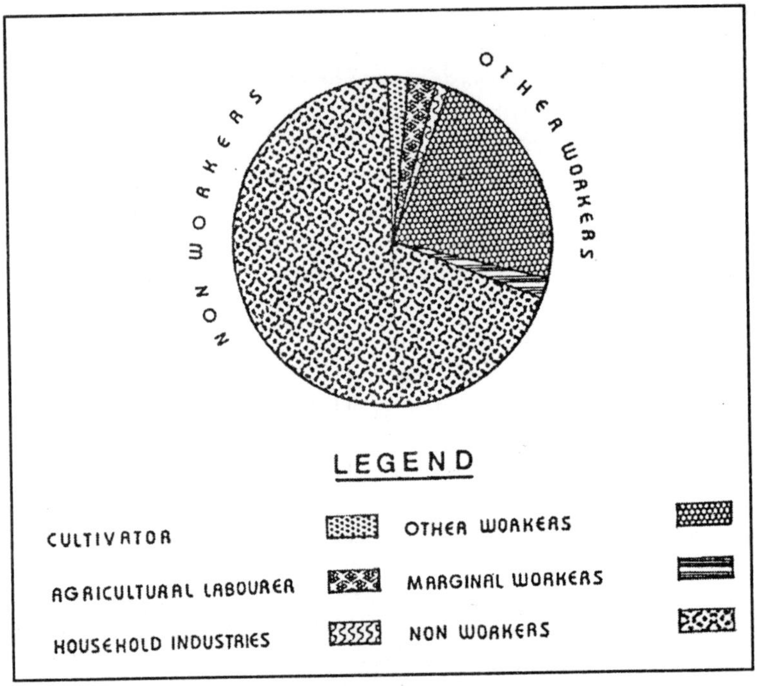

FIG. 8.16 : *Occupational Structure of Hirakud Town, 1981*

During 1951, the total population of the area was 6, 559. With fresh immigrants from outside due to construction of the Dam and industrial development, the population figure was increased to 15, 040 during 1971. Within 1951-61 ths rate of total increase of population was 2034 and 6,447 in 1961-71, i.e., 203 persons per year between 1951-61 and 645 per year between 1961-71 respectively. The growth rate during 1951-61 was 31 per cent and 74 per cent in 1961-71.

Table 8.12 : Variation and Growth Rate of Hirakud Town

Year	Total Population	Decadal variation	Density per sq.km.	Decadal growth rate (in p.c.)
1951	6,559	—	506	—
1961	8,593	+ 2,034	664	31
1971	15,040	+ 6,447	1,161	74.9
1981	21,701	+ 6,661	1,676	54.82
1991	23,830	+ 2,129	1,840	44.45

In 1971 the density of population of the town was 1,161 per sq.km. against 506 in 1951. Obviously this high growth rate was experienced by enormous flow of people from different parts of the country. The new industrial establishments of Hirakud Dam are the main employers. The construction of Hirakud Dam Project is responsible at the first stage of town development.

1961 : The town is divided into seven wards for smooth administration. The following table shows the population characteristics in each ward.

Table 8.13 : Population Characteristics of Hirakud by Wards, 1961

No. of Wards	No. of occupied residential houses	No. of total persons	Percentage of the total population of the town
I.	205	863	10.00
II.	580	2, 046	23. 71
III.	300	765	8.90
IV.	302	902	10.48
V.	242	980	11.40
VI.	639	1,945	22.63
VII.	321	1,092	12.70

Ward No. II & VI are important wherein 46 per cent of the total population are clustered.

During 1961, the sex-ratio of the town was 658 female per 1000 males. The low percentage of female is due to the fact that in the industrial establishment male workers are very much higher than the female. More than 19 per cent of the total population of the town belongs to Scheduled Caste and Scheduled Tribe category. More than 50 per cent of the total population were literate at that time.

About 45.19 per cent of the total population of Hirakud were workers and 54.81 per cent non-workers. Female worker were negligible only 451 out of 3,884 workers. In 1961, 31.95 per cent of the total workers were engaged in different services and at that period majority of the workers were engaged in the town. The labouring class came from the adjoining villages like Mahabatpur, Jamadarpali, Tihura and Nua Tihura. They did not stay within the town but would prefer to cultivate the green fields around the developed areas of the town. Besides these a large number of population are

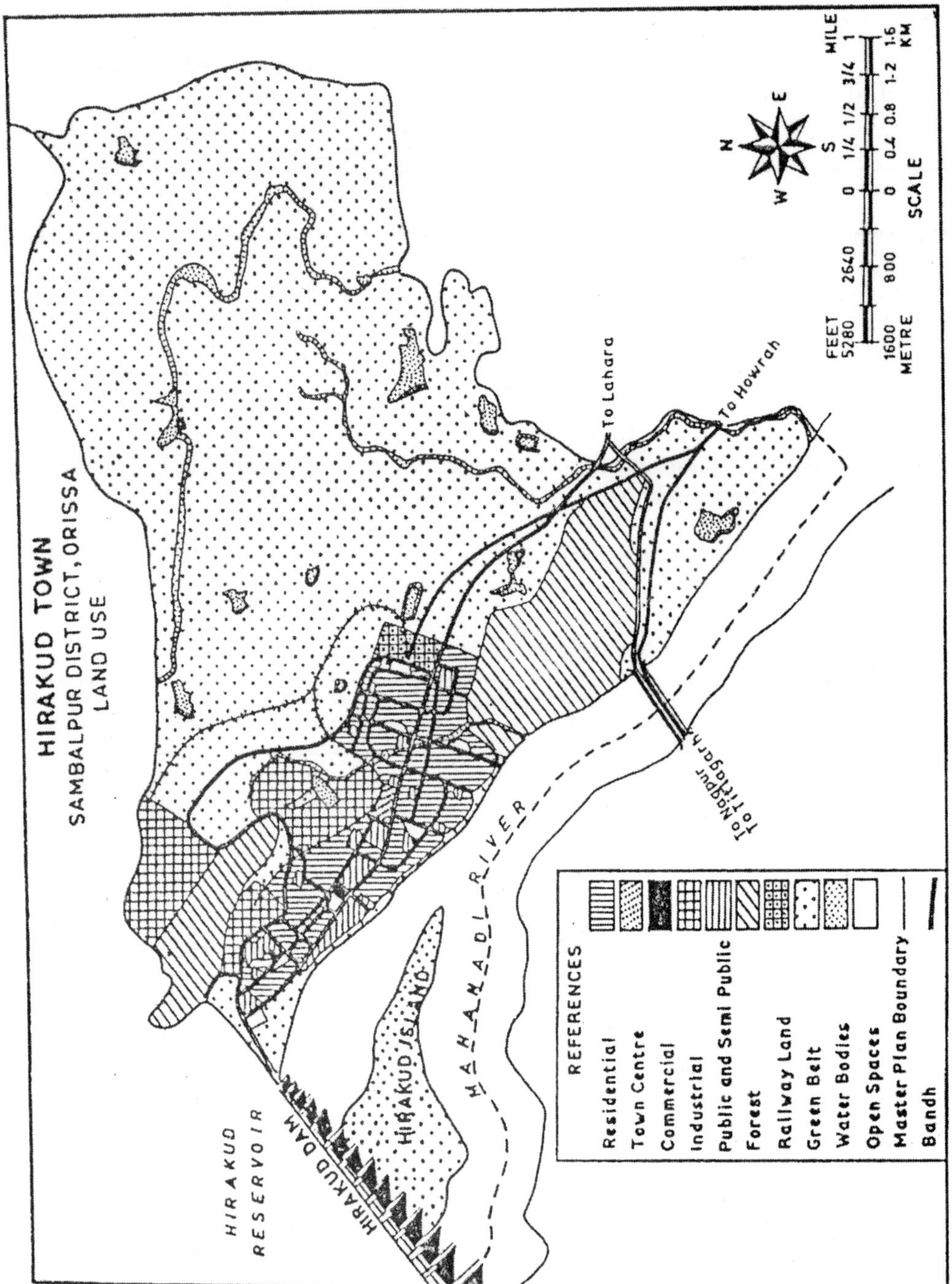

FIG. 8.16 (a) : *Morphology of Hirakud Town*

PLATE 17 : Indal Community Centre, Hirakud Town.

PLATE 18 : Hirakud Town, Market Centre.

engaged in manufacturing industry. With rapid expansion of industries the percentage of workers is gradually increasing though the recent data are not available at present.

Functional Analysis

From the detailed analysis of occupational structure of the town it seems that the town is by-functional in character. From the field observation it is realised that industry is the backbone of the town. Cultivators and agricultural labourers are limited in number. Workers engaged in mining and quarrying are also negligible. More than 26 per cent of the total workers are engaged in household and miscellaneous manufacturing industries. The household industries are bidi-making, pottery, weaving, mat-making. The manufacturing industries are (1) Indian Aluminium Co., (2) Aluminium Industries (3) The Hira Cable Factory (4) Hirakud Workshop (5) Hirakud Automobile Workshop and (6) Hirakud Ice and Glass Mills. The future prospect of industrial development of the area is bright as the rich mineral base of the state could be effectively utilised with the help of hydro-electricity from Hirakud Project, which is supplying the bulk quantity of power to the Rourkela Steel Plant at present.

Above 32 per cent of working class were engaged in construction according to 1961 Census. They were engaged mainly in different constructional works of the Hirakud Dam Circle, factories and above all Hirakud township.

Commercial activity of the town gives very little impression of the occupational structure of the town. The present township has encouraged only few people in this activity. Transport, storage and communication play insignificant role amongst the town dwellers, only 2 per cent is engaged in this activity. Some of the inhabitants are engaged in transporting to the town dwellers and transporting goods to the outer areas by taxi, rickshaw, tempos. Nearly 32 per cent of the town workers are engaged in different services. Recently there is a tendency of chance in the occupational character of the people. Within ten years the construction of Hirakud Dam Project has been completed and different activities of the project is going in full motion.

Proposed Landuse

Hirakud is now being developed as an industrial town under Notified Area Council. In future, Hirakud will be an important industrial town in Orissa. The establishment of a planned township is considered for the development of economy of the district in particular and the State as a whole. A Master Plan of the town was prepared by the Government of Orissa in consultation with the NAC. In the plan there are provisions for systematic development of industries, residential areas commercial and infrastructural facilities for the setting of ancillary and small scale industries. It is understood that this proposed landuse plan may have to undergo some modifications when detail layouts and designs would be prepared. An area of about 604 ha. will be covered in the Master Plan for industrial purpose, the township and essential services.

Work in the Hirakud town has already been started as per the Master Plan. The land is being developed gradually for different purposes according to the orientation of the left dyke of the Reservoir. The elevation of the dyke is sufficiently high to prevent flooding which may occur due to heavy rain in the Mahanadi Catchment due to cyclone.

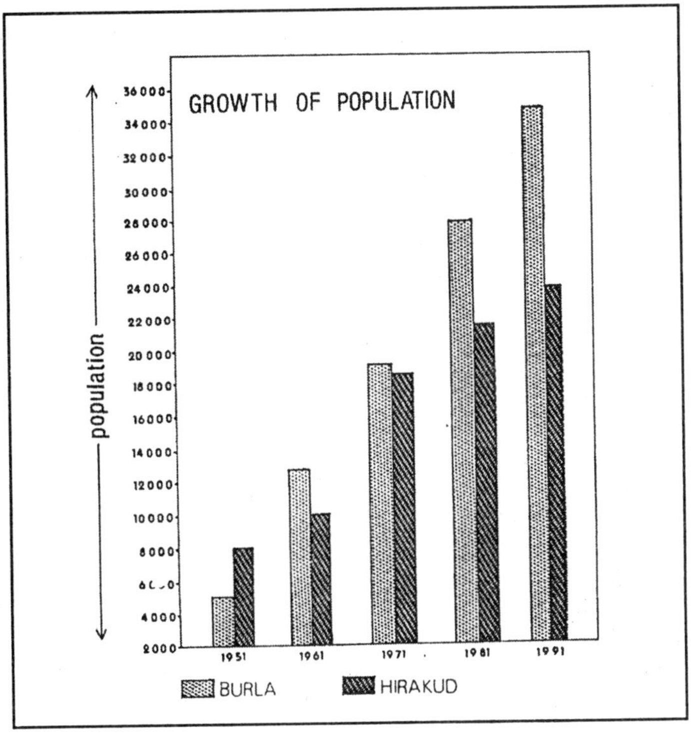

FIG. 8.17 : *Population Growth of Burla and Hirakud Towns*

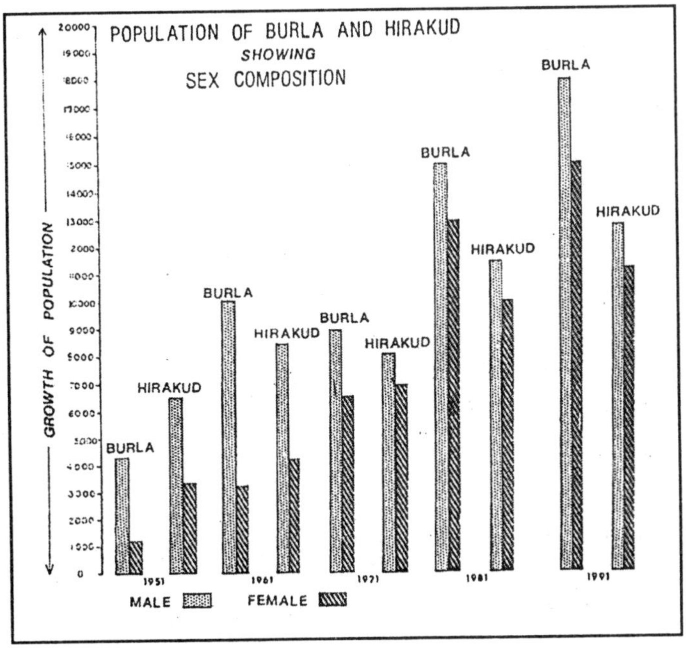

FIG. 8.18 : *Sex Composition of Burla and Hirakud Towns*

PLATE 19 : Indian Aluminium Factory, Hirakud Town

PLATE 20 : Hirakud Industrial Work, Hirakud Town

PLATE 21 : Hirakud Airport — At Extreme Point of Left Dyke

PLATE 22 : Runway of Hirakud Airport

Table 8.14 : Proposed Landuse of Hirakud Town

Categories	Percentage of developed land
Residential	40.19
Commercial	1. 40
Industrial	29. 61
Public and Semi-public	13.46
Open space for future development	4.08
Roads	7.91
Railway	3.35

Communication

The transport system is more or less the same as of Burla. The long distance passengers generally prefer to travel in train from Hirakud Railway Station. There is direct connection with Sambalpur town by road. Taxis and Tempos are available. The town is also connected with adjoining villages by motorable metalled and unmetalled roads.

Comparative Study

The two towns Burla and Hirakud are separated from each other by the Mahanadi River and the distance between them is 5 km. But in future these towns should not be merged into one as the bed of the Mahanadi is to be kept open always for the discharge of the excess water of the reservoir or any type of calamities which may occur due to which the river will flow its own course. Only a bridge can unite them to act as complementary to each other. Both the towns are the outcome of Hirakud Dam Project but there are basic functional differences between the two. Service is the predominant function of Burla. More than 62 per cent of the total working class of Burla is engaged in different services. But Hirakud is developing as an industrial town, where nearly 54 per cent of the total workers are employed in manufacturing industries and in construction.

9

CASE STUDIES OF SELECTED VILLAGES

General Outline

Introduction of irrigation by Hirakud Reservoir has changed the rural scenario of Sambalpur District to a considerable extent. In the Canal Command Area, two giant canals – one Bargarh and other Sason and their distributaries have made Sambalpur District lovely green from its previous bleak landscape which looked like dusty brown. From enlarged air-photographs and satellite imagery, the canal system is studied and the green fields of paddy are clearly identified. It is gathered from field investigation that three crops of paddy are harvested in the irrigated areas. Multiple cropping of paddy throughout the Canal Command Area has improved the economic conditions of the local people to a substantial extent. To monitor such changes some villages have been selected here for detailed investigation, in addition to geographical information system derived from aerial photographs and satellite imagery. Six villages are chosen near the dykes like Jamadarpali and Balbaspur and some near National Highway No. 6 (Calcutta-Bombay) such as Gurbhaga, Ladarpali, Suktapali and Kurla. From this field study, it has been ascertained that irrigation water does not reach all parts of the arable land due to variation of micro-relief. As for example, at Jamadarpali about half of the arable land is not getting irrigation facility of the Sambalpur distributery system as the northern portion of the village is situated at a higher elevation. It is not possible to provide canal water from canal network at elevated portions of the village. The provision of lift irrigation from the distributaries is however prohibited. Further it is also ascertained that harvested rice can not be sold at remunerative prices due to lack of good communication system. So six villages portray contradictory socio-economic pictures though located within the same Canal Command Area. Detailed investigation of the area reveals that Gurbhaga is most prosperous village of the area. The people from Andhra Pradesh, particularly the Telegus took the first advantage to cultivate the land using irrigation water and reap bumper harvest. On the contrary, the local people were initially reluctant to use irrigation water due to superstition and failed to avail of the opportunities provided and thus, continue to remain poor.

JAMADARAPALI V ILLAGE

Jamadarpali is a village of moderate size lying in the underdeveloped tract of the country in the Sadar Subdivision of Sambalpur District in Orissa.

Location

The village lies within the jurisdiction of Hirakud Police Station and is situated approximately in 21°33' N Latitude and 83°59' E Longitude. It is bounded in the north by the Hirakud Reservoir, in the south by Lamdungri Reserve Forest and in the east by Pandripali and Hirapali village and in the west by Larbanga and Mohramunda village.

The village is about 10 km. from Sambalpur town. The village was once situated on the southern bend of the Ib river at a distance of about 2.5 km. from it. But, after the construction of Hirakud Dam the Ib river is now merged into the Reservoir and the village is now situated on the South Eastern end of the Left Dyke of the Reservoir at a distance of about half a kilometre from the Left Dyke. This village is at a distance of about 6 km. from Hirakud Railway Station as well as about the same distance from Sason Railway Station. But there is no transport link between this village and the Railway Station. But in dry season one can reach the village by car.

So it is approachable only by foot and bullock-cart from Hirakud Police Station. In fair weather the unmetalled road can be used for jeeps and trucks and can easily be reached from Sambalpur town or Sason Railway Station.

Physical Setting

From the left bank dyke of the Hirakud Dam, the village appears picturesque, with groups of hamlets with sal trees and bamboo thickets. The unmetalled road with red murrum cover imparts the picturesque beauty of the village. There is a small *nala* on the western side of the village which runs from north-west to south-east directions. In the neighbourhood there are isolated small ridges on the north and on the eastern side of the village. Part of the northern ridges is now submerged within the reservoir with their highest portions standing above the water. The 180 m. contour line passes through the village with general slope towards the south. Red and yellow soils are the characteristics of this area with occasional lateritic patches. The surface is mostly undulating making agriculture difficult and canal irrigation impossible. No stream passes through the village nor is there any irrigation canal.

Climate

The village is located far from the sea coast. The climate is characterised by hot summer and cool winter. The vast stretches of waters of the Hirakud reservoir keeps the temperature cooler by 3°C. During April to June, the heat is intense. The cool weather starts from November and ends in February. These months are practically rainless. The much wanted monsoon bursts in the month of June and ends in the middle of September with occasional showers for a month thereafter with much variation of rainfall. The village had no rain recording station. Figures obtained from Meteorological Office at Sambalpur indicates an annual average of 2, 000 mm. of rainfall.

Source of Water

The main source of water for the village are the Tanks known as the 'Bandha' and the other type known as 'Kota'. These tanks are made mainly for the purpose of irrigation. But none of them have been renovated for several decades and are therefore lying silted up and are not of much use. The villagers take bath in these tanks, wash their clothes and also drink water from them. There is no separate well for drinking purpose. As the soil is shallow and rock based, the construction of wells

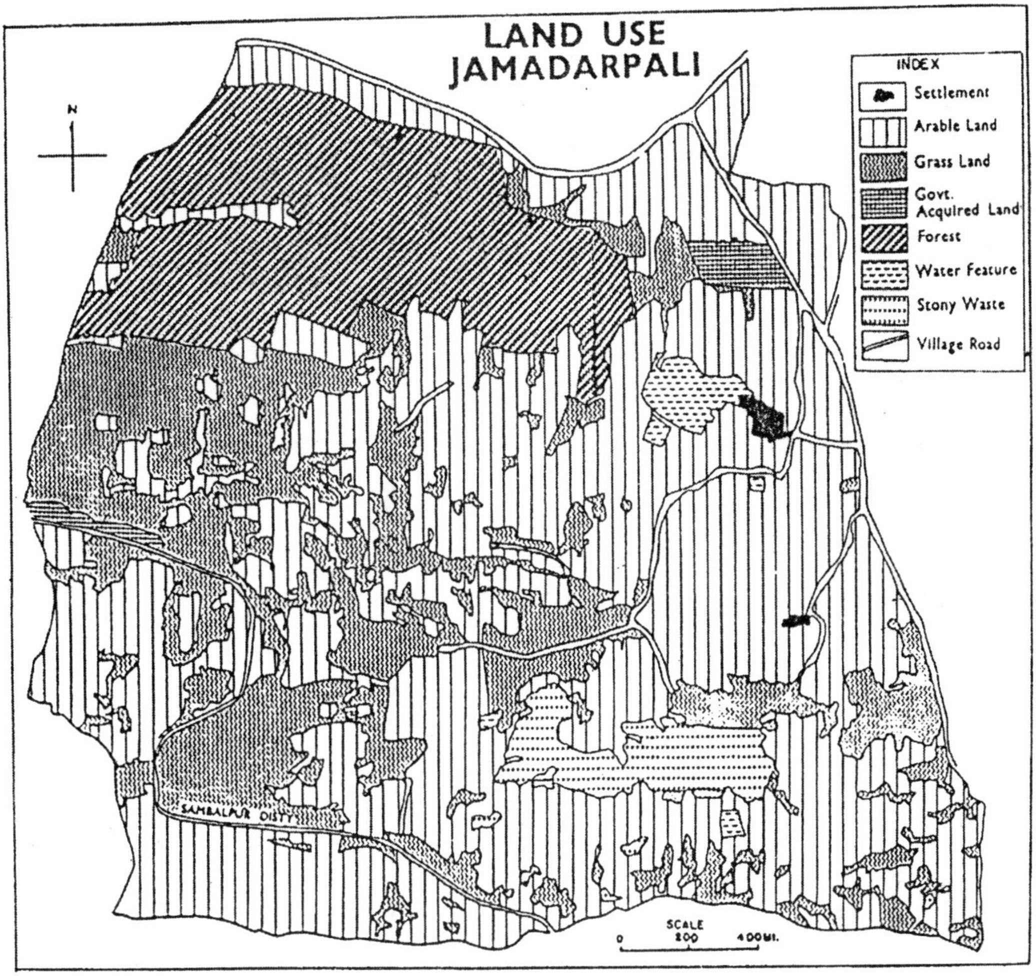

FIG. 9.1 : *Landuse Map of Jamadarpali Village*

become very costly. The small nala which was constructed some decade back mainly for the purpose of irrigation gets dried-up for the greater part of the year keeping the villagers helplessly relying on monsoon rain for their agricultural production.

Flora and Fauna

The village and its surroundings are rich in flora and fauna. The climate and soil are suited for the growth of various species of vegetation. The principle varieties of trees are — banyan, pipal, mango, date-palm, sal, mahua, kendu, dhaura, kusum and palas. There are also bamboo groves and jungles around the village specially in the northern side. The villagers are constantly cutting down the trees for the purpose of cooking and to light fire in winter thereby reducing the forest resources.

PLATE 23 : Jamadarpali Village Post Office

PLATE 24 : Shaby Condition of Jamadarpali Village devoid of Canal Irrigation

PLATE 25 : Sason Canal coming out of Hirakud Reservoir.

PLATE 26 : Sason Canal Down Stream.

Population

The village has a population of 1011 as per 1981 Census. The population of 1951 were 745 persons, of which 392 were males and 353 females. This is an increase of 4.2 per cent in 10 years. The present population comprises of 502 males and 509 females. This shows not only the fair proportion of women in the total population but also their growth at a rate much faster than that of man of this village. It is still a remarkable fact that while woman population has increased at a rate of 10.5 per cent in 10 years the male population has actually decreased at a rate of 1.3 per cent during the same period. Women are dominant in these villages not only numerically but also in all social and domestic walks of life.

All the villagers are Hindu by religion. A majority of the people are Scheduled Tribes numbering 699 of which 352 are males and 347 females. There are 67 Scheduled Caste people of which 34 are male and 33 female. Amongst 1011 persons only 217 persons are literates of which 199 are male and 18 female. Literacy has not been increased to any appreciable degree. There is only one Primary School in this village which is in a very bad shape. The number of school going children is limited. There are only 14 per cent literates as compared to 40.9 per cent literates in the whole of Hirakud Police Station and still much less than Sadar Subdivision (26.6 per cent). This accounts for the backwardness of the village both socially and economically.

It appears that the village with an area of 5.6 sq.km. has a population of 1011 which works out to be a density of 180 per sq.km. This is less than the average density of the District which is 223 or even the average density of the State which is 292. The density of Hirakud Police Station is 194 per sq.km. which is more than Jamadarpali. It, therefore, appears that this abnormal rise in density of population might be due to new settlers who have migrated from the neighbouring villages, now submerged within the Hirakud reservoir. Moreover, many workers who come in this village during the construction of Hirakud Dam, prefer to stay back after the construction was over.

As the density of population is high, this village probably will not be able to support any further rise of population in future. Unless the resources of the village are increased by various development measures, any further rise in the population will lead to further poverty and misery to the local people.

Houses

The village has 219 houses and equal number of households for 1011 persons. During the last 10 years, the number of houses has increased from 134 to 219 although the total number of persons increased were 234 and the households increased by 85 in number. The village has clustered type of settlements with long rows of houses on both sides of a narrow village road. The road is wide enough to allow two carts to pass side by side. The village has a number of hamlets situated within close distance to each another. They present an orderly look. Each house has one or two rooms with a rectangular verandah.

House Type

Most of the houses are of *Kutchcha* type consisting of one or two rooms, which are generally rectangular in shape. They have no separate compound but a few well-to-do persons have more commodious houses with three or even four rooms. Most of the houses have a wide courtyard in the back side of the house for storing grains or stacking straw.

Construction of houses is simple and economical. The materials used are stone pieces, bamboo and wooden poles, thatch grass, date-palm leaves and strings. The walls are made up of split bamboo and twigs and often with stone pieces and earth. A coating of mud is plastered on both sides of the walls. The walls are constructed up to a maximum height of 3.5 m. over which the roof is built. The roofs are made up of thin stone slabs, locally procured, being mixed with thatched straw or tiles or with date-palm leaves with split bamboo. There are some brick-walled houses in this village belonging to some rich cultivator classes. The roofs have mostly two simple slopes in the usual manner protruding much beyond the walls to protect them from rain. The rooms have mostly no windows and having only one door to each room. Therefore, the rooms are usually dark even during the day time. The walls of the houses are coated with mud and are given cow dung wash. The floor is levelled and is finally plastered with cow dung and liquid mud.

The approximate cost of construction of a typical house of peasant's family is Rs. 150 and that of a poor Kamar is Rs. 50.

Dress

The villagers usually use clothing of cheap and coarse varieties restricted almost to bare necessity. Men and women wear handloom fabrics woven by Bhulias. Women usually prefer *sarees* of smaller size. Red colour border is popular among the village women. The *sarees* are worn in a manner so that they hang up to a level just below the knees with no tendency to flutter in movements. This type has practical utility and is congenial for household work.

Man invariably wear *dhotis* either of handloom or of mill made. The cultivators usually wear *banians* and occasionally readymade shirts to cover the upper portion of the body. Children up to 6-7 years remain naked. School going boys put on half-pants while the girls use small size *sarees*.

Food and Drink

The staple food of the villagers is rice which is the largest single item in their diet. Rice is taken throughout the year in whatever quantity or form.

The poor are content with a stomachful of torani which is the local name of water allowed to stand over a quantity of cooked rice throughout the night turning a little sour by process of fermentation. Those who are well-to-do take '*pakhal*' that is water rice of the preceding evening with chillies or onions or condiments or with '*sag*' or vegetables.

The people usually take an adequate quantity of freshly cooked rice which is taken with vegetables, *curry* and *dal*, chillies, tamarind, onion and salt. Evening meal is almost the same as mid-day meal. It thus, appears that the food taken by the villagers is ill balanced, fat, milk and its products meat, fish and eggs are seldom taken. This lead to higher mortality amongst the aged people. That is why there are few aged people in the general population of the village. The male population who work more are declining in number.

Economic Resources

The economic life of the villagers is very simple. They are running their life on marginal economy. A very high percentage of population of this village is wholly engaged in the production of food. Even women and children are engaged in this job. Those villagers are wholly dependent on agriculture for their subsistence. But, although, they depend wholly on agriculture, they have not improved the technique of cultivation. The method of cultivation still continues to be simple and old fashioned leading to low production of crops. Though the village is situated at the periphery of the Hirakud Reservoir the villagers do not get any advantage of it. No irrigation canal passes through this village. Even the Sambalpur distributory canal which flow through the western side of this village at a distance of about 1.5 km. is of no use to the villagers, as its water cannot be utilised in any way for irrigation. It is the irony of fate that though the village is situated within the command area of the Dam, the village is completely deprived of the Reservoir water. Sason canal which flows at a remarkably rapid speed from the Reservoir at a distance of about 2.5 km. from this village thereby depriving the villagers from utilizing the canal water for their agricultural production.

About half of the total area of this village has been brought under cultivation. Of the remaining areas about one third is covered with grass pasture lands and the rest covered with jungle.

It is, therefore, evident that agricultural production is far below the needs of the entire population of the village. The villagers therefore regularly go to the local market for purchase of rice for their daily needs.

All the cultivated lands are therefore rainfed. As stated above the village has no irrigation facilities though it is situated within the command area of the Reservoir. The village has seven irrigation tanks out of which 6 are known as ' *Kotas*' and only one known as ' *Bandha*' . Those irrigation tanks generally get dried up during those parts of the year when the water is most needed for irrigating the fields for the production of paddy. Most of these tanks are silted-up and remain dry except on rainy season. The villagers use these tanks to meet their domestic requirements including drinking water. The water is only 90 to 120 m. deep along the bund side. The other sides are completely dry. No attempt has been made to improve these tanks either by the villagers or by the government. The cultivators of this village, therefore, entirely, depend on monsoon rains which is completely seasonal and uncertain in nature. Added to this, the rainfall is unevenly distributed throughout the year thereby bringing disaster to cultivation. Hence, the cultivators of this village get only one crop a year depending on natural rain.

The agricultural lands are divided into three classes, *viz.* , Bahal, Mal and At. Most fertile first class wet lands are locally known as ' Bahal' . Bahal and Mal lands grow paddy and At lands grow Rabi crops including vegetables. The pattern of agricultural practice is very old and almost in the same lines as their forefathers used to do. No modern method or skill is employed in the operation. The villagers even do not know the use of improved variety of seeds or use of chemicals or fertilizer for their agricultural production. Hence, the cultivators hardly get only one crop a year and that too depends on monsoons. If the monsoon fails, the crop fails and disaster comes to this village.

Economic Activities

The villagers can be classified into different occupational activities based on their agricultural and other economic pursuits. Out of 1011 (1981) people of this village, 147 are cultivators of which all are male. Only 6 people are engaged as agricultural labourers. About 397 persons are non-workers of which 189 are male and 208 female only. Eighteen people are engaged in cottage industries including Bidi making and other services, 164 persons are engaged on household industries of which 81 are male and 83 female, only one persons is engaged in mining.

There is only one Primary School in this village. The number of school going children are few. One post office functions in this village, being operated by two persons one Post Master and the other one Postman.

The most popular occupation of this village is therefore agriculture. One-fourth of the population of this village is engaged in this occupation. This is mainly due to scarcity of agricultural land, most of which have become unfit for agriculture due to non-availability of water. The rest are fallow lands, grass lands and scrub lands. It was also noticed that most of the agricultural lands of this village has undulating surface on which it is difficult to hold water for the purpose of agriculture.

To give the villagers off-season relief, the government provides them part-time or even whole-time jobs in the construction of new roads or repairing the existing one on the Dam side.

Another household industry is the making of earthern pots of various types and shape. The village has no weaver or Bhulia. These facts show the poor economic condition of the village.

Occupational Mobility

A high degree of conservatism is noticed in the matter of occupation. These villagers prefer to stick to their traditional occupation. None is willing to undertake a voluntary change which require a considerable amount of risk and strain and certain amount of ambition and spirit. The villagers have hardly displayed such character or temperament. Only one persons is engaged in mining. He has changed his profession from ancestral cultivation to mining. A few person is seen engaged in fishing on the shallow waters of the Reservoir to sell their catch to the nearby weekly Balbaspur market village or at the Hirakud town. Very few persons have opted to take service in the industries at Hirakud town which is not very far from this village and has a high remmunerative value.

This shows that the villagers are very conservative in their change of occupation and are satisfied with their meagre income drawn out of their farm lands.

LADARPALI VILLAGE

Location

Ladarpali is a village of Attabira Police Station of Sambalpur District, Orissa. It is situated on the Bargarh-Sambalpur Section of the National Highway No. 6. the distance of this village from Bargarh is about 17 km. west and from Attabira it is about 5 km. east. It lies roughly at the crossing of 83°46' E longitude and 21°22' N latitude.

Buses and taxis running on this road stop at this point. Small tea and snack shops are found here serving the persons waiting for the transport. Road transport is much prominent here as the village is not connected with railway line.

Landscape

Ladarpali village presents a flat topography with occasional big trees here and there. The shape of this village is almost like a rectangle with a little tapering in the south. The average height of this village is 165 m. The western portion of this village is a bit higher in comparison to the east. Almost at the centre of the village , there is a big pond attached with a ' bundh' . The contour embankment of that pond is on the eastern side. Almost all the plots of this village are under irrigation either by canals or by tanks. The Kota mentioned above plays an important part in the irrigation system of this village. Comparatively higher plots which do not receive irrigation water are transformed to grass land. There are few grass land patches in this village. Otherwise the whole village is properly cultivated. The NH 6 passes through the northern boundary of this village. Babebira distributary flows out from the Attabira branch canal. It forms the southern boundary of this village.

Climate

The village has got no special characteristics so far as climate is taken into consideration. We have discussed general climatic condition of this region when we studied the Gurbhaga village of Attabira Police Station. Here also monsoon rain is well distributed from the month of June to the middle of September. The high percentage of moisture content in the air makes the weather very sultry and oppresive, specially during rainy season. From the month of November the temperature begins to drop and December becomes the coldest month of the year. The day temperature in the hottest month (*i.e.* May) rises approximately to 41°C. The mean minimum temperature in the coldest month falls to 12°C approximately. The average annual rainfall is about 1500 to 1750 mm. 90 per cent of which is received during monsoon months.

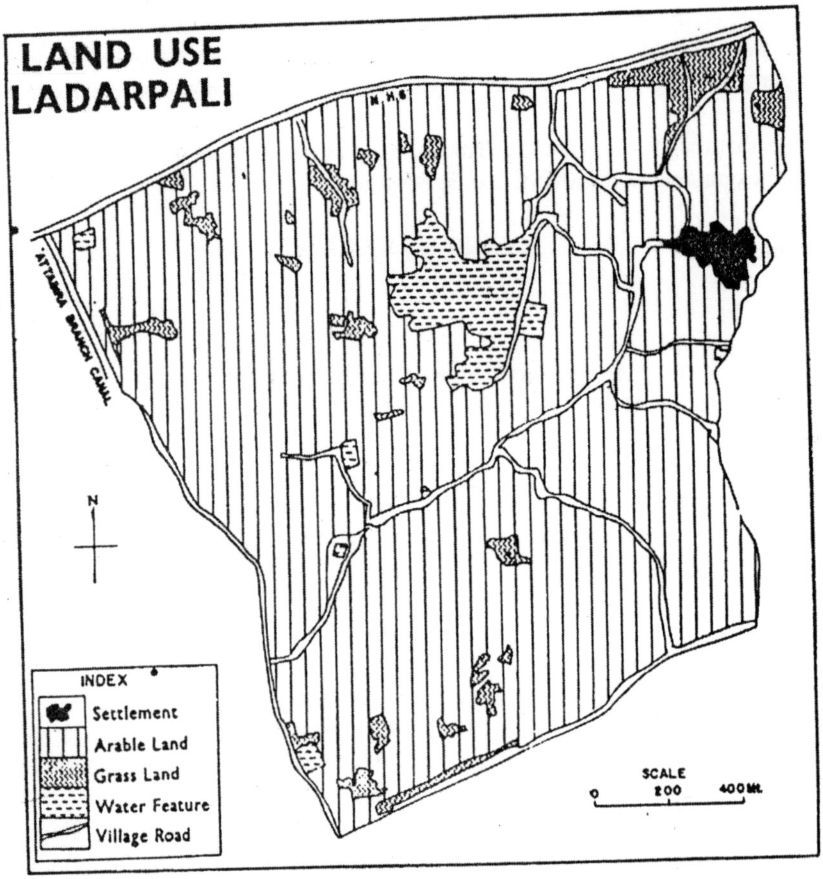

FIG. 9.2 : *Landuse Map of Ladarpali Town*

People

Ladarpali has got on area of 2.60 sq.km. with a population of 1688 out of which 864 are male and 824 are female according to 1981 census. The village is a very small having only one *Basti* (Hamlet) situated in the eastern portion of this village. Though small, it is one of the most densely populated villages of the Attabira Police Station. The density of 649 persons per sq. km. is much higher than that of the district average of 105 persons per sq.km. respectively. The density of 270 persons per sq.km. in the year 1951 was equally higher compared with the then density of 94 persons per sq.km. of the State. The increase in the density in the year 1961 is commensurate with the growth of population during the decade. The decade of 1951-61 was a period of economic prosperity. The completion of Hirakud Dam Project has created a gross irrigation potentiality. The growth of population in the decade has risen to an extent of 19 per cent which was much higher in comparison to the corresponding figure of 15.9 per cent of the district average.

The village has got 432 number of Scheduled Caste persons and 196 number of Scheduled Tribe persons and they contribute 17 and 13 per cent respectively to the total population. The number of females in both the categories are equal to those of the males.

Housing

The number of houses and households occupied by the residents of this village are 337 and 366 respectively. The villagers have settled together in compact form on a higher land leaving the low lying areas for cultivation. There is a solitary *Basti* (Hamlet) where the houses are rectangular in shape and constructed in the traditional style. Where is no opening of the house along the road side other than the main door. The courtyards are in the backside of the houses. These courtyards however do not allow much light on air inside the houses. The rooms are usually very dark.

Majority of the houses are built of mud walls and thatched roofs. But brick walls are also common. At the entry of the village, there are a number of houses built with bricks, cement, etc., concrete rods are also used for the roofs. But most of the houses use tiles and .thatches for the roofs. In the midst of these hutments, there is an open space, meant for social get-together, panchayat meetings, marriages, etc.

Irrigation

Ladarpali is a well irrigated village, where most of the lands have been transformed into prosperous farm lands. There are patches of grass land, where canal water cannot reach due to height. Being within the canal command area this village enjoys the full benefit of the irrigation system. The crossing of Attabira branch canals and National Highway forms the northwestern corner of this village. The main distributary which supplies water to the fields is the Babibira distributary, which forms the southern boundary of the village. This distributary receives water from Attabira canal.

Before the introduction of canal water, tanks used to supply water to the adjoining fields. But now the tanks are not of much use to the villagers. They take bath and some cases drink water from these tanks .

An attempt is made to cultivate wheat in the comparatively high land of this area. These uplands are locally known as 'at' lands. As canal water cannot reach this land, cultivation of wheat is practised here, which requires less water. The total number of cultivators and agricultural labourer of this village is 463 which comprises about 31 per cent of the total population and 67 per cent of the total working population of the village. This shows that the village is mostly dependent on agricultural economy. The canal system of Hirakud has considerably increased the facilities of irrigation here.

But in comparison to the local people, the Andhra farmers have proved themselves competent enough in the field of agricultural activities. They are hard working and are accustomed to modern techniques of cultivation. Some of the farmers of this community are much more prosperous than the local farmers. Now the local people are gradually utilising effectively the canal water and fertilizers, etc. and trying to compete with the Andhra farmers.

Economic Activities

The working population of this village can be grouped into five broad industrial categories. These categories are taken from the nine major categories mentioned in the census. The five categories are (1) Cultivator, (2) Agricultural labour, (3) Household industries, (4) Trade and Commerce, and (5) Other services as per 1981 Census.

The total working population of this village is 651 persons of which male 527 and female 124, which contributes 46 per cent of the total population of the village. The number of persons falling under each category is as follows . 193 persons as cultivator, 270 persons as agricultural labourer, 59 persons as household industries, 129 persons are engaged in other services.

The highest numbers of persons of this village are grouped as cultivators. They comprises 28 per cent of the total population. Thus, we can easily say that this village is a full-fledged agricultural village.

The next important occupation is cotton weaving in handlooms. About 15 establishments are there engaged in cotton weaving. But the quality of the cloth is not very good. The weavers locally known as ' *bhulias*' who produce coarser variety of cloth — mostly used by the local people. Some local carpenters are also there but their number is not very high. About four families are engaged in carpentry work. Very few are engaged in bidi-making.

Collection of 'Mahua' flowers and preparation of country liquor from it is also an important occupation. But this always forms a side occupation of the villagers.

Collection of 'Kendu' leaves and selling it to the government agents form an important economic activity amongst most of the persons either male or female who are otherwise not engaged in any other services.

Rice is the staple cereal of this village. Rice from this area is taken to Bargarh market which is a big and good market for agricultural products.

The number of persons in the non-working category of this village is 1017 out of which 558 persons are males and 459 persons are females. Other than dependents and infants the females attending to the household industries are also included in this category. Male members who are otherwise engaged in household duties are also included.

One thing has been noticed by us that though the economy of the villages we have visited have remarkably changed, the villagers do not want to agree that whenever they were asked about the change of economy, they usually denied that. Though, they agree to the fact that due to availability of canal water they are not going to accept the fact that certain amount of economic prosperity they are enjoying now. It may be due to the fact that they have not learnt how to improve their living standard by breaking the cultural stagnation or to be free from the traditional prejudices with which they are influenced. They are leading the same type of life they used to lead before the introduction of the canal water irrigated agriculture.

PLATE 27 : Sason Canal of Hirakud Reservoir showing Data Plaque.

PLATE 28 : Gurbhaga Distributory coming out of Bargarh Canal.

GURBHAGA VILLAGE

Gurbhaga is a village in Attabira Police Station which has been, much transformed after adopting irrigated agricultural system, being benefitted from the Hirakud Dam. This is one of the most prosperous villages of the said thana.

Location

The village Gurbhaga lies almost exactly at the crossing of 83°50' E long. and 21°25' N lat. The Bargarh-Sambalpur section of the NH 6 passes through this village at the middle. The village is about 26 km. east of the Sub-divisional town Bargarh. One can easily identify Gurbhaga because a name plate made of wood and iron pole has been put up prominently on the road side. Buses and taxis running on this road stop at this point. A number of small shops are found on either sides of the road. Though Sambalpur-Titlagarh railway line passes through these villages, the station is at Attabira, which is far away from the village. Very few travel by train to commute to the market town of Bargarh or Sambalpur to the east. They prefer to travel by bus.

Landscape

The average height of this village is 165 m. with little variations of 1.5 m. to 3 m. in places. The general slope is from north-east to south-west with an elevated portion at the centre. That is why the Gurbhaga distributary canal alongwith its field channels flow in southerly direction, irrigating the adjoining areas and leaving back the comparatively higher northern portions. Thus, we get here a clear demarcation line made by the irrigation channel — the south of which is being irrigated properly by the canal water, whereas, the higher land of the northern portion does not receive any canal water and has become the pasture land of the village.

Previously most of the village was covered with scrub jungle specially the upland areas. But with the construction of distributary and field channels, the picture has been changed. Most of the lands are now properly irrigated with the help of a network of water channels. Only the uplands, locally known as 'at' land, are left aside because of the elevation of that area.

The shape of Gurbhaga village is more or less longitudinal and has become narrow in its north and south end. The main canal that is Gurbhaga Distributary Canal passes through this village first in the north-south direction, then from north-west to south-east and again in a north-south direction. There are two other canals coming out from this Gurbhaga Distributary; one is Tupapara Minor and another is Nilipali Minor. Tupapara Minor flows to the west of the Gurbhaga Distributary. But both of them cover very little area that also in northern portion of this village. As we have already pointed out that the northern portion of this village is comparatively high, main distributary has to cut the track forming ravine of average depth. Thus the canal passes about one and half km. distance in the north-west south-east direction cutting a gorge of about 3-4.5 m. As there is no possibility of construction of any field channel here, the area is not getting any irrigation particularly to the left side of the main distributary.

Except the area mentioned above that is the 'at' land of the northeastern portion of the village, it has been observed during field survey that the canal water was flowing in full swing and was entering in the paddy fields through the field channels and the growth of paddy was good in the month of April, 1973. The cultivators usually get a bumper crop.

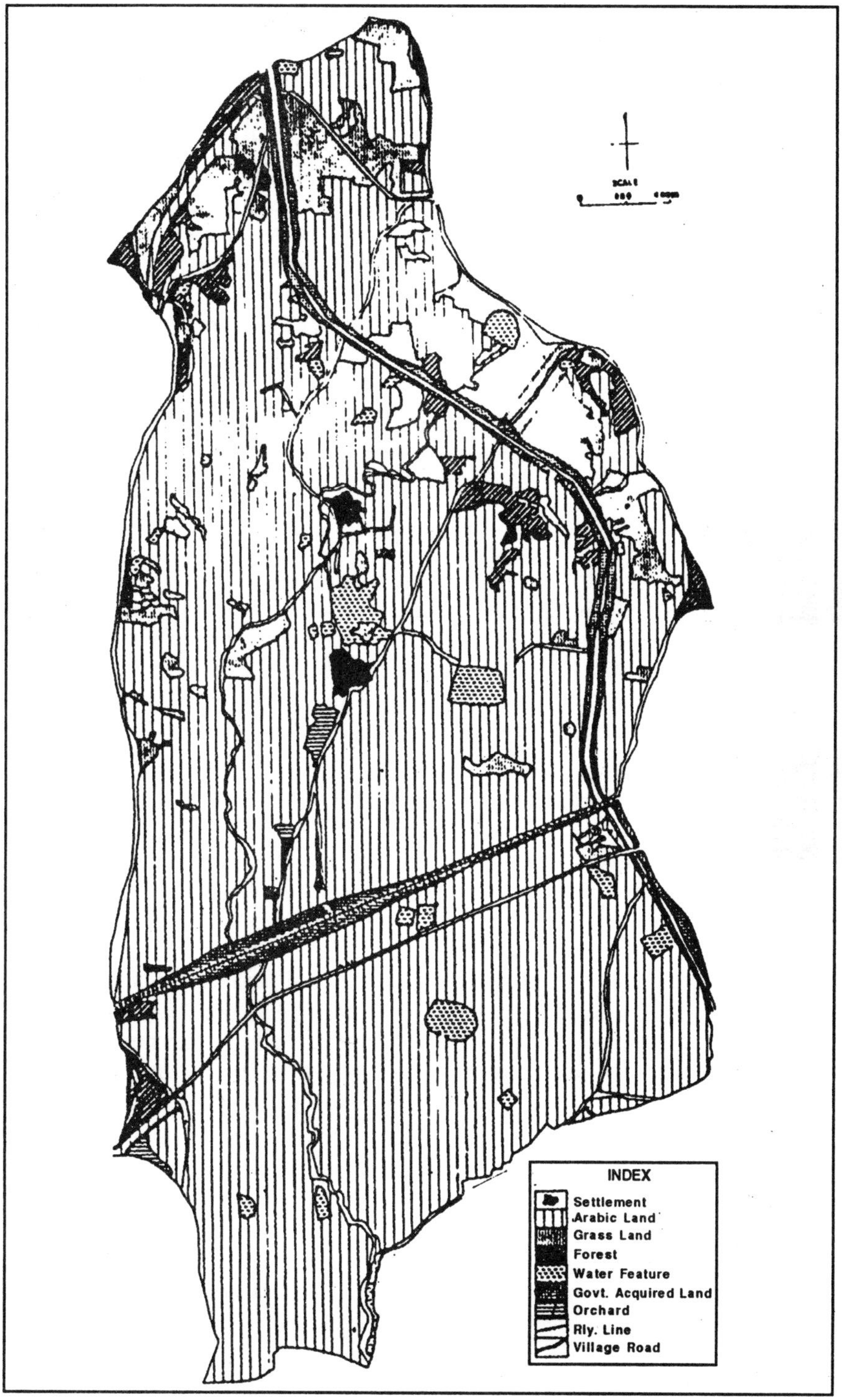

FIG. 9.3 : Landuse Map of Gurbhaga Village.

Climate

In this village, the climate is identical to that of the district – that is a hot summer and well distributed rain during the south-west monsoon season. Temperature rises from early March steadily till May, which is the hottest month of the year. The mean maximum temperature of May is 41.6°C. The onset of monsoon is from middle of June, and lasts up to September. Though the rain brings some relief to the heat, high percentage of relative humidity of the air causes the weather to be hot and oppressive. From November, the day temperature also begins to drop rapidly and December and January are usually the coldest months of the year. The mean daily maximum temperature of winter is 28.5°C and mean daily minimum temperature is 12.1°C.

The average annual rainfall of this area is about 1500 mm. About 90 per cent of the annual rainfall is received in the monsoon months. Relative humidity is high during south-west monsoon season. The existance of Hirakud reservoir, a vast water body of about 777 sq. km., has influenced the climate of this region as a whole. Though the extremity of the climate has been checked to some extent, the percentage of relative humidity has been greatly increased.

People

Gurbhaga village covers about 104 sq.km. area and has got a population of 4,105 persons in 1981 of which 2087 are males and 2018 are females according to 1981 census. Broadly speaking the number of males are equal to that of females. In 1951, the proportion of female population was identical of the total population of 1251. In 1951, 614 people were males and 637 were females. In 1951 population was 1251 and in 1981 the figure is 4105. The figure is much higher in comparison to the Police Station as well as the District average. The density of population of this village is 395 persons per sq.km., whereas, the District figure is 105 and State figure is 169 only. Its placement within the command area of Hirakud Canal System, existence of a network of field channels which irrigate all possible cultivable lands, improved technique of agricultural activities — all these helped this village to attract a large number of persons from outlying areas.

The village has got two clusters consisting of about 902 households. One is Gurbhaga and another is Kendapali. The distance between the two *mahallas* is about one kilometre only and both are situated on the central part of the village.

There are two schools in this village, one is primary and the other is middle school. The total number of literate persons here is 1539, out of which 1055 persons are males and 484 persons are females in 1981. The percentage of literate in this village comes about 31 per cent of the total population. The percentage of literacy of the police station is about 27 and of the District is 23 only. Accordingly, in 1951 the percentage of literate people was 28 approximately. The most important thing to mark here is the increase in the percentage number of literate female population. During 1951, there were only, 46 females who were considered literate and within a span of ten years *i.e.* in 1961, the figure recorded an increase of 112 that is from 7 per cent to 13.5 per cent. This indicates a general inclination to the female education of this village.

The total number of Scheduled Caste and Scheduled Tribe population of this village were 878 and 936 respectively in 1981. Most of them are engaged in agricultural activities.

Housing

The village has got 338 houses and 400 households. Most of the houses have mudwalls and thatched roofs. Mud huts are in majority and few houses have brick built. Most of the roofs of the mud wall houses are thatched. But roofs of corrugated tin sheets, as well as concrete roofs with brick walls are also found in very few houses. Only well-to-do farmers may venture to build their houses like that. The villagers live together in a comparatively raised land leaving aside the agricultural fields, which are located in low lying areas. The settlement patch is called *Basti* locally, and each *Basti* is provided with one grocery shop, place for community meeting, at least one well for drinking water, etc. This village has got two such *Bastis viz*. Gurbhaga and Kendpali. Each *Basti* is provided with a big pond (kota) attached with a *bandh.* A place for social function of community meeting is left aside by the villagers in every village. A rectangular or square space is left aside and the houses are constructed encircling this piece of land. During festival time the villagers bring tree leaves from nearby jungle and make a shed with the help of 'sal' poles (bamboos are also used) and those tree leaves. Thus a very soothing and green cool shed is formed under which people gather and do work. These sheds are very helpful during the hot days of summer months and last for about a couple of weeks.

Usually the houses are built along the side of the village roads in a linear pattern. The main entry of the house is through a big door placed on the road side. But other than this main door there is no other opening in any form *i.e.* windows etc. on the road side. The courtyards etc. of the houses are always constructed at the back of the house. Most of the houses are having two roofs made of thatch or sometimes country tiles or sometimes a combination of both. In most of the cases, the roof is extended sufficiently to cover a long front open varandah constructed alongwith the house on the road side. Almost all the households posses equipments like country cots, locally made of bamboo frames with coir fibre base, plough etc.

The main occupation of this village is agriculture which is mostly done by canal irrigation. Previously tank was most popular. Large number of village tanks are the prominent features of this village. Even today these tanks (there are three types of tanks namely *kota, Mundas* and *Bandh* according to their sizes, depths and embankments, etc.) supply water to a minimum of 2 hac. to maximum of 40 hac. according to their sizes of course. These tanks are also very useful where canal water is not available. After the opening of the Hirakud Canal system, the importance of tanks is gradually decreasing as the canal irrigation is gradually becoming popular. In Gurbhaga village, there are a number of tanks and Kotas but nowadays they do not play any prominent role in irrigation. Most of the tanks are gradually drying up.

Irrigation

It has been stated that this village is gradually sloping towards the south-west. Thus, the higher portion of this village which covers the north-east portion cannot enjoy the facility of canal water and whatever irrigation is done only by *Kotas* and *Bunds*. The rest of the village is irrigated most effectively with a network of field channels connected with the main distributary — the Gurbhaga Distributary Canal. Introduction of chemical fertilizers, improved techniques of cultivation and high yielding varieties of seeds have enabled the farmers here to become economically well-off in comparison to the farmers settled outside the command area.

Economic Activities

Out of 4105 persons of this village 1217 persons are classified as 'workers' and 412 persons 'non-workers'. By implication 'workers' include all persons working as well as the working children irrespective of their earnings. The working population constitutes 40 per cent of the total population of the village. The main occupation of the villagers is agriculture. Of the total population 414 persons (out of which 402 persons are males and 12 females) are cultivators and 917 persons (out of which 528 persons are males and 389 persons are females) are agricultural labourers during 1981. Thus, about 59 per cent of the working population is engaged in agricultural activities. In can be said that majority of the people of this village is dependant on agricultural occupation.

Thirty-seven persons are engaged in household industries, in making earthen wares, textile products, country-made cigarettes, shoe etc. A number of families are engaged in handloom weaving. They produce coarser varieties of cloth, mainly for local consumers. Some families are engaged in manufacturing earthen wares and earthen potteries. This profession comes next to agricultural activities when number of persons engaged are taken into consideration. The open courtyard is used for drying materials made of soft clay. Subsequently, the dried earthen wares are burnt in closed ovens, using fire woods from nearby forests. The next important non-agricultural occupation is manufacturing of baskets and other materials from bamboos, canes, leaves,etc. Bamboos are available from nearby forests and the materials made out of those are easily marketable either locally or to the big markets of Bargarh and Sambalpur. Some local cobblers were also found.

The trade in this village centres revolve round the agriculture products. Rice is the main export item of this village. Collection of kendu leaf is a subsidiary occupation of the villagers. Manufacturing of 'Bidi' is gaining importance now a days and each and every member of the family are engaged in this business. A brick field of quite a big size has also been set up in the eastern bank of Gurbhaga Distributary across the railway line. This brick field is situated on the 'At' land which is devoid of any canal water. But for the work of the factory in the brick field the canal authority have allowed them to lift water from the Distributary.

KURLA VILLAGE

Kurla is situated within the Canal Command Area and is one of the prosperous villages of the Bargarh P.S. of Sambalpur District, Orissa.

Location

The village lies almost at the meeting point of the Matwali Jhor and the Danta river. The Matwali Jhor, a tributary to the Danta, forms the eastern boundary of the village. The Danta river flows in the north-west to south-east direction and forms the southern limit of the village. Kurla lies roughly at the crossing of 83°41' 30" E long. and 21°23' N lat. This village cannot be easily approached, as it neither lies on the National Highway nor is connected by any good motorable road.

Landscape

The village presents a typical landscape of this region that is a flat topography with scattered trees. Almost all the plots of this village are cultivated. Only a little area of the north-west is comparatively higher and has become the pasture land of this village. The average height of this village is about 165m. The middle portion is slightly elevated and the land is gradually sloping towards the Danta river in the south-west and the Matwali Jhor towards south-east. The northwestern portion of the village meant for pasture has at average elevation of 166 m. and is sloping towards south. A little area adjoining to this pasture land is covered with a very thin forest — locally known as village woodlot.

In 1981, the village had a total population of 1768 of which 896 were males and 872 females. The

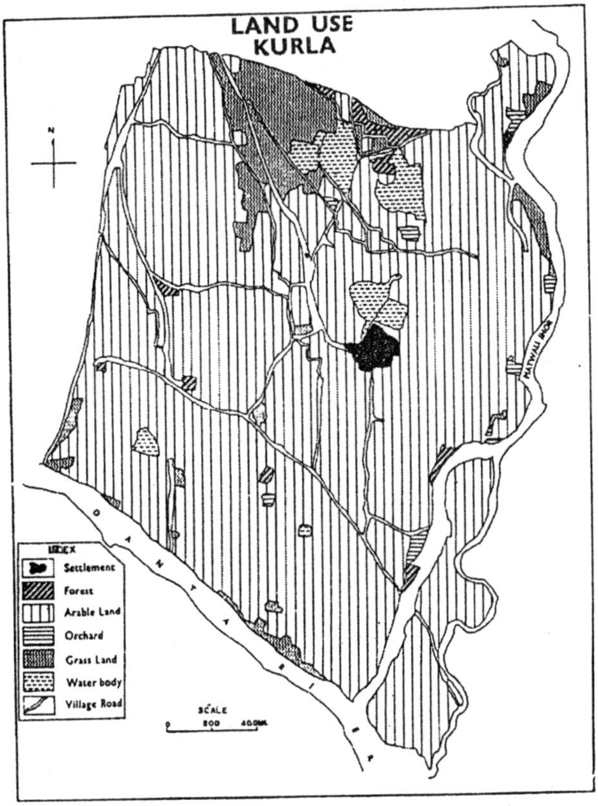

FIG. 9.4 : *Landuse Map of Kurla Village.*

village has got a Scheduled Caste population of 497 persons and Scheduled Tribe population of 183 persons only. They contribute 28 and 10 per cent to the total population of the village respectively.

In 1981, the level of literacy in the village was 23.5 per cent. The figure is almost identical to other villages of the District, and slightly above the average level of the State. There is one primary school in this village.

Housing

The total number of houses and households occupied by the residents of this village are 318 and 369 respectively. The houses are, built in an irregular cluster, winding lanes and blind alleys. Most of the houses are built along the lanes and they face each other. Very few houses are built of bricks. Thatched roofs are common. Tiled roofs can also be noticed. Palm leaves and bamboo sticks are sometimes used for the construction of roofs. The materials are found in abundance in the village woodlot. The walls are sometimes made of bricks of dried mud with bamboo sticks. In both the cases, both sides of the walls are plustered with mud. In case of *kuccha* houses, the owners themselves contribute unskilled labourers. The completion of the hut does not require any skilled labour. As usual the courtyards are mostly found in the back sides of the houses. Usually they are used for drying grains. In a weaver's house the courtyard is used for spreading, dying and also designing the cloth to be woven.

Economic Activities

Persons classified according to their chief occupations are as follows : Cultivators 236, Agricultural labourers 447, household industries 21 and other workers 32. The number of total working population of this village is 736 out of which 527 are males and 209 females. The number of persons engaged in agricultural activities is 683 which contributes 83.7 per cent of the total working population. This indicates that the village is out and out an agricultural village and the main occupation of the villagers is agriculture.

Production of edible oil and cotton weaving are worth mentioning. About eight establishments are engaged in pressing oil from ground nuts. There are four units engaged in weaving cloths. The quality of the cloth is of coarser variety. There is a cobbler shop in this village.

Besides agriculture, the villagers have got a number of subsidiary occupations which sometimes help them particularly in years when monsoons fail. These are helpful to the villagers who reside at a comparatively higher land ('At' land) where there are no irrigation facilities. Kendu leaf collection, bidi making, collection of mahua flowers etc. are some of these occupations.

The non-workers of this village are students, women engaged in household works, children and aged persons. Out of 1031 non-workers 369 are males and 662 females. They contribute 21 and 37 per cent. In many households women combine household duties with some other items of wage earning work like bidi-making, kendu leaf collecting, mahua flower collecting etc, .

Irrigation

Kurla village lies within Hirakud Canal Command Area System. Bargarh Police Station is practically divided by the Bargarh canal into two portions, one is the western portion which is outside the Canal Command Area and other is the eastern portion within the command area. Kurla lies in the east of Bargarh Canal.

Proper distribution of field channels in this region and supply of water to the fields through these channels have helped the farmers to irrigate their lands intimately, thereby to reap two to three crops in a year, a phenomenon which was beyond imagination before the construction of Hirakud Dam. The whole area was once covered with scrub and very limited area was put under farming. But now the landuse pattern of this region has changed radically.

Though the villages within the Canal Command Area have gained much prosperity, the net benefit accrued to the local people is negligible as many people have no land and alternate occupation is lacking. But people living within the Command Area are better off than those without any irrigation facilities.

BALBASPUR VILLAGE

Balbaspur is a fairly large village, about 3 km. from Jamadarpali village. The village is situated just on the left dyke of the Hirakud Dam to the eastern end of it. It is under Sason P.S., of Sambalpur district of Orissa.

Location

It is situated approximately at 21°34' N lat. and 84° E long. It is bounded on the north and west by the Hirakud Reservoir and on the south by Pandrpali and Harapali villages and on the east by the uplands and scrub lands of Sanganpali village.

The village is about 12 km. from Sambalpur town but lies within 6 km. from Sason Railway Station. It is situated within 5 km. from Sambalpur-Jharsuguda State Highway. But the village is situated outside the bus route. It is however connected by jeepable road in dry season. There is an airstrip on the northeastern side at a distance of about 1 km. from this village.

It appears that the village was once situated on the south-eastern bank of the Ib river where it takes a westerly turn to join the Mahanadi. There was ferry services from this village for going to Mangalpur, Lakhanpur and other villages situated on the northern bank of the Ib river. But this has since been stopped after the construction of the Hirakud Dam, thereby submerging the villages on the northern side of the Ib river. This village being the weekly market place is well connected with the neighbouring villages by unmetalled murrum road enabling jeeps and trucks to pass through them during dry season.

Physical Frame

This village has the same physiography as Jamadarpali except that the land of this village is not so much undulating as the former. The average height of the village is 182 m. above sea level. The general slope is to the south. The eastern side of the village is an upland and two or three small streams rise from this upland flows in the southern and eastern directions. A small stream (*nala*) passes through the village from west to east. But it gets dried up in hot weather. The village has a small patch of village woodlot on the southeastern side as well as on the north-eastern side.

The soils are mostly red and yellow with gneissic base. Exposed gneissic rocks occur as out-crops. The soil appears to be very fertile as it is situated on the bank of the Ib river, one of the main tributary of the Mahanadi.

Climate

Being situated on the edge of the vast water-body, the village is neither very hot in summer nor very cool in winter. But the laterite and sandy soil and exposed rocks make the climate hot in summer. Hence in the months of April, May and June the climate is hot and sticky and the winters are cool. Usually the monsoon breaks in the month of June and ends in the month of September. Being very close to Jamadarpali village, Balbaspur has the same amount of rainfall as that of Jamadarpali.

Source of Water

Though it is situated on the edge of the Hirakud Reservoir the villagers cannot fully utilize the water of the Reservoir to meet their irrigation requirements. The main source of water therefore are tanks, locally known as *Bunds* and *Kotas*. There are 9 such bunds and only one *Kota* in this village and the villagers irrigate their fields from these tanks. But most of these tanks are in a very bad shape. They are mostly silted-up and do not hold much water and get dried up during summer. There is also one big masonary well just at the entrance of the village constructed by the Government about 17 years back when this was termed as a model village bringing in new settlers from the neighbouring submerged ones. The villagers use the water of the well mainly for drinking and cooking purposes. The tanks are used for domestic purpose including for bathing, washing clothes and cattles.

There is a small stream which passes from west to east bifurcating the village in two halves. This is mainly used for the purpose of irrigating the fields. But it gets dried up during dry season as it is not

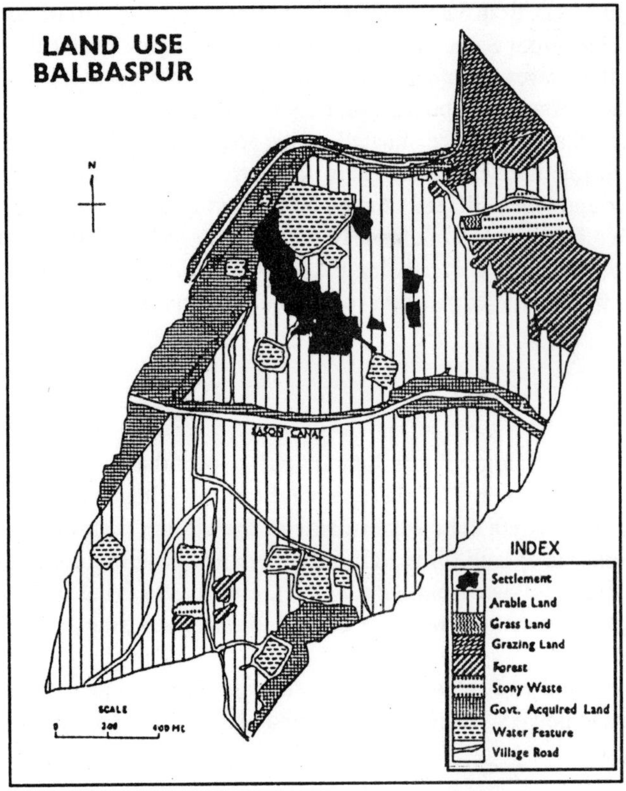

FIG. 9.5 : *Landuse Map of Balbaspur Village*

linked up with any of the irrigation canals. The southern portion of this village gets the advantage of lift irrigation from the swiftly flowing Sason Canal. As the village has a general slope towards the south no field canal could be constructed from the said canal to irrigate its fields except by lift irrigation system at some places. This lift irrigation is practised unofficially.

Population

The village had a population of about 964 persons in 1981, out of which 477 were males and 487 females. The population has increased from 705 to 964 during 1951-81, thereby increasing at a rate of 12.19 persons in 10 years which is a negligible increase. In 1951 before the construction of the Dam, the total population of the village was 705, of which 327 were males and 378 females. Out of this 705 persons, 505 belonged to agricultural class and 200 to non-agricultural classes. Ten years back there were 190 occupied houses and equal numbers of households for 705 persons. But now the number of occupied houses have swelled up to 206 and number of households to 208. In 1981 the number of houses is 222 and household is also 222. Total number of literate persons have also increased during these 10 years. In 1981 total number of literates were 326, of which 233 were males and 93 females. The village has a high percentage of literacy. There are about 36 per cent literates in this village, compared to 26.6 per cent in the whole of the sub-division and 23.9 per cent in Sason P.S. This shows that this village is well advanced socially and culturally than the neighbouring villages.

There are 407 Scheduled Caste people and 163 Scheduled Tribes living in this village in 1981. The villagers are predominantly Hindus by religion.

Most significant features of this village is the decrease of agricultural class. In 1981 only 92 persons were cultivators of which 88 persons were males and 4 females. There are 95 agricultural labourers of which 86 were males and 9 females in 1981. Like Jamadarpali, the important characteristic of this village is that the women are dominant not only numerically but also in all walks of life.

The density of population is very high when compared to the density of the subdivision as well as in the Police Station. The density of population of this village is about 372 per sq.km. which is much higher than the average density of population for the whole of Orissa which is about 169 and about 163 for the whole of Sason P.S.

Houses

The Balbaspur village had 222 occupied houses and 222 households for 964 persons in 1981. The houses have increased from 190 to 222 during these 10 years although the total number of population increased from 705 to 914 an increase of 209 persons during last 30 years. This may be due to settlers from the neighbouring submerged villages after the construction of Hirakud Dam. This village gets all possible help from the Government in cash and kind so that a model village could be set up for the affected villagers.

The village has typical linear type of settlement with long rows of houses along both sides of the village road. The houses are elongated from west to east direction with southern side generally closed having even no window for the room. But eastern sides are open to allow the sun and light to enter the houses. There is an open verandah facing the road. The village has a long and wide road built by the government. Each house has one or two rooms with a wide courtyard inside for storing grains or stacking straw. Most of these houses have *dhenki* --- a large wooden morter and pestle for husking paddy. This device is kept at the entrance corridor of the house. No house has windows facing the village road so as to maintain privacy for the female members of the family.

Most of the houses are built of mud like those of Jamadarpali but some have tiled roofs. They have no separate compound. Each house has at least one bed room, one kitchen and one verandah. A few well-to-do persons have two or three rooms with a wide courtyard to stack straw.

The construction of houses is simple and economical. The materials used are the same as those of Jamadarpali. The only difference is that the houses of this village are having more orderly look than Jamadarpali. The houses are well maintained and are frequently repaired. This may be due to the fact that the villagers here are economically well off than that of Jamadarpali village.

The village has a recreation club centre built by the government. The villagers usually meet here during their leisure for recreation.

Economic Resources

The life of the villagers are very simple as that Jamadarpali village. About 30 years back 505 persons out of the total of 705 were engaged in agricultural production. But now 92 persons are cultivators and 95 persons are agricultural labourers. Although the villagers are wholly engaged in agriculture, their method of cultivation still remains old-fashioned in the same line as their forefathers. They have not improved in any way. Hence, their agricultural production has not increased to any remarkable degree.

The village having been situated within the Canal Command Area of Hirakud reservoir does not get full benefit of irrigation from the canals. Some land near the Sason Canal gets the advantage of lift irrigation. The major portion of agricultural land entirely depends on rain. Irrigation tanks known as bunds and kotas are not of much use as the water of these tanks get dried up during the dry season, leaving the villagers completely dependent on monsoon rain.

There are only 193 ha. of land, all of which are not cultivated. Such a small amount of land cannot adequately feed the villagers. Hence, the villagers have to go the local market to purchase their daily needs. This shows that the village is not self-sufficient in the production of rice. The food production is much below the normal needs of the villagers. Only one crop is grown on most of the fields except in those portions which get the benefit of the Sason Canal where two crops are grown in a year. Hence, on the whole the economic condition of the villagers are not very bright and the villagers live on marginal economy. Main crop grown in this village is rice. Vegetables are grown on at lands alongwith some Rabi crops. But that too is far below the requirements of the villagers. About 2000 kg. of paddy is grown per hectare of Bahal or first class wet land situated on the canal side and about 600 kg. per hectare on mal lands outside the canal area. About 250 kg. (0.4 mt) of Rabi crops per hac. are grown on comparatively high or 'at' lands.

Economic Activities

The village can be classified into different occupational groups based on their agricultural and economic activities. Out of 964 persons, about 92 persons are cultivators, of which 88 are males and 4 females. Out of 95 agricultural labourers engaged in the production of food, 86 are males and 9 females. These labourers work on contract basis from morning to evening. They get certain amount of paddy either on daily basis or on monthly basis. Other important economic activities of the village is handloom weaving. The reels for this industries are supplied by the cooperatives. About 91 persons out of 964 are engaged in this household industry, of which 34 are males and 57 females. Only 9 persons are engaged in mining. About 70 persons are engaged in various industrial units at Hirakud town situated at a distance of about 7 km. from this village. These industries are Hira Aluminium, Hira Cables and other small industries. Some persons are engaged in trade and commerce, selling handloom products to the local market and some others are engaged in selling fish, netted from the Hirakud Reservoir.

All these activities show that the village is on marginal economy. The population of this village is not self-sufficient so far as the production of food is concerned. The agricultural land is not enough to feed the entire population of this village. Hence, the villagers have to purchase their food often from the local market. There is acute shortage of agricultural land in this village, compared to its population. There is only 175 hectares of land of which all are not agricultural land. Hence the agricultural production is far below the need of the people. The villagers, therefore, spend larger share of their income on food. Their standard of living is low. But this village seems to be more prosperous and economically better of than the neighbouring Jamadarpali village. The houses are built with the same material locally available. They appear better and are well maintained. These villagers are better dressed and are better fed than Jamadarpali. These are due mainly to several factors, *viz.* presence of handloom industries, production of double crops and large number of persons engaged in industries at Hirakud Town. But no such factors are present at Jamadarpali village, which is deprived of canal water irrigation except a few hectares leading to low production of crops. Hence Jamadarpali is economically and socially much more backward than that of Balbaspur, Balbaspur being well connected with the neighbouring villages, the villagers here are able to sell their products goods at the weekly market of their own village.

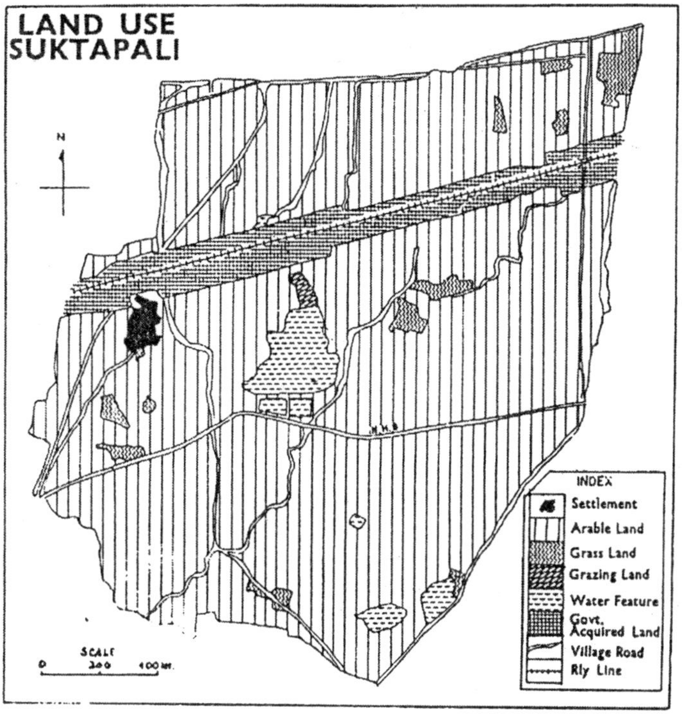

FIG. 9.6 : *Landuse Map of Suklapali Village*

SUKTAPALI VILLAGE

Introduction

Suktapali is a small village well situated on the NH 6 not very far from Bargarh town under Attabira P.S. Sambalpur district, Orissa. It lies mid-way between Bargarh and Attabira town and is a prosperous tract of the Hirakud ayacut region well fed by its canal water.

Suktapali village is situated at 21°22' 30" N lat. and 83°44' E long. It is within 10 km from Attabira on NH 6. It is situated about 3 km east of the Danta river, a main tributary of the Jira river. The village is well situated and can easily be reached by any means of transport in all seasons. A wide unmetalled road passes through this village making it easily accessible throughout the year.

In the north the village is bounded by the Matwali Jhor, a tributary of the Danta river. On the east lies Manapara village and on the south by NH 6 and on the west by the Danta river. The village is conveniently situated within the Ayacut region of the Hirakud Dam. There is a railway station close to Attabira Branch Canal on the eastern side of the village. This village is well connected by railway line connecting Titlagarh and Jharsuguda stations.

Physical Layout

The village has a flat topography, having an average elevation of 170 m. above the mean sea sevel. The village has an average slope towards the south-west and most of the streams and canals run in that direction. The most important canal known as Kalapani Minor passes through this village from east to south-west direction. The soil is red laterite and are made very fertile under the cover of canal irrigation. Patches of grasslands occur on the northeastern part of the village. The land being flat the construction of canals is easy and cheap leading to high production of agricultural crops without much artificial manuring. But the soil is not much thick having granite base.

Climate

The climate of the village is characterised by a hot dry summer. It receives well distributed rain from south-west monsoon. The temperature begins to rise from the month of March till May which is the hottest month of the year. The mean daily temperature is about 41°C during summer months. The heat in May and early part of June is very tiresome. The weather becomes much oppressive due to high moisture content of air. The nights are comparatively cooler due to the presence of water in the paddy fields around the village. From the month of November the daily temperature begins to drop considerably. December is usually the coldest month of the year with maximum temperature around 25°C and minimum temperature 12°C.

The average rainfall of this village is good. The rainfall is about 1500-1750 mm a year 'from South-West monsoon'. Relative humidity is very high specially during the rainy season. Winters are generally dry and cool. The village gets wide-spread rain and sometimes strong winds from the south-west monsoon.

Source of Water

This village is conveniently situated within Hirakud Canal Command Area and is fed by canal water throughout the year. One such canal known as Kalapani Minor originating from Debahal Distributary passes through this village in east-west direction dividing the village into two halves. Many field channels from this canal passes through this village watering all the paddy fields. There is another canal known as Sukutapali Minor which passes through this village. This shows that there is no shortage of water for the villagers to irrigate their paddy fields throughout the year. Hence the village does not have to depend on monsoon rains which are uncertain and ill-distributed. There are a number of field channels and field outlets well distributed to cover up all the crop lands of this village enabling the villagers to raise their crops twice or even thrice in a year. All these distributaries and minor canals are fed by voluminous Attabira Branch Canal which has originated from the Bargarh Canal.

In addition to these there are number of irrigation tanks locally known as *Bunds* and *Kotas* to irrigate comparatively higher 'At' lands where irrigation by field channels are difficult. There are 8 such tanks. These tanks are also used by the people for domestic purpose. The villagers take their baths in these tanks and wash their clothes and bathe their cattle. There is a well in this village and the villagers take water from this well mainly for drinking and cooking. Most of these tanks are rain fed and they get dried up in dry season. They are not well maintained and therefore get silted up.

Population

In 1981, this village had a total population of 740 persons of which 358 were males and 382 females. In 1951 this village had only 308 persons of which 143 were males and 165 females. There were 97 occupied houses and equal number of households in 1951. But during the 10 years period, the number of occupied houses have decreased to 91 and number of households to 94. In 1981, the number of houses was 152 and households 152. There were about 218 persons in agricultural occupations of which 172 were males and 46 females. The number of agricultural labourers in this village is 133. This shows that there is an increase in agricultural labourers. These labourers have no land of their own. They work on the farm lands of others either on daily or on monthly wage basis.

The number of Scheduled Castes residing in this village was 254. There were 73 Scheduled Tribes in 1981.

About 12 persons were engaged in household industries locally known as *Bhubas* or weavers. Out of these 7 persons were males and 5 females. About 5 persons were engaged in trade and commerce. The number of other workers was only 26.

In 1951, there were 84 literate persons in this village. But now the figure stands at 309, of which 213 are males and 96 females. The village is well advanced in literacy while the percentage of literacy in the whole of Bargarh Subdivision is 21.2 per cent and that of Attabira Police Station 26.6 per cent, the literacy of this village is about 35 per cent, though there is only one Primary School in this village.

The population of this village has not increased much when compared to the increase of population in the whole Subdivision. While the increase of population in the whole Subdivision, is 4.27 per cent, the population increase of this village has increased at the rate of about 1.7 per cent per year.

Density of population of this village is 122 per sq.km. This is slightly more than the average density in the whole Subdivision which is about 108 per sq.km. This shows that the village is able to maintain even larger population as the village is well fed by net work of canals from the Hirakud Reservoir.

Houses

The village had 152 occupied houses and 152 households for 740 persons in 1951. In 1951, the number of occupied houses were 97 and an equal number of households for 308 persons. Although the population has increased from 308 to 740 during these 30 years, the number of occupied houses had increased from 97 to 152 and the number of households from 97 to 152. Suktapali is a compact village with linear type settlements. The houses are separated by a long and narrow village roads and lanes. These houses are rectangular in shape and are generally open on the eastern and southern side with a courtyard spacious enough, generally on the back portion of a house mainly for the storage of paddy as well as for stacking the straw. Most of these houses have a wooden morter and pestle for husking *(dhenki)* paddy. These *dhenkis* are kept either in the corridor of the house or on the verandah. But the use of *dhenki* is becoming less due to husking of rice in machine. The village roads and lanes are wide enough to allow the carts to pass side by side.

The houses of the villagers are rectangular in shape and are generally have opening on the southern and eastern side to allow the light and air to enter into the house. These houses have in general a courtyard on the back side of the house to allow for the storage of paddy snd stacking of straw. Most of these houses have 'dhenki' for the husking of paddy erected either inside the room or in the verandah.

House Type

Most of these houses like the other two villages are of *kuccha* type comprising of one or two rooms with a rectangular type verandah on the entrance of each house. Some houses have wide verandah on the back side of the house. The houses in general have no windows. Hence, they are generally dark even during the day time.

The construction of the houses of the villagers are very simple and economical. They generally use the materials available locally. The materials used are the same as those used at Jamadarpali and Balbaspur village. The only difference is that the houses of this village are more commodious and well maintained. There are a number of 3 or even 4 roomed houses with wide courtyards for receiving grains. There are only a few *pukka* houses built by rich cultivators. But they are not well maintained by the villagers.

Economic Activities

The village can be classified into different occupational groups based on their economic and agricultural activities. The number belonging to cultivation class was 85 only out of a total population of 740 of this, all are male. Alongwith this there are 133 agricultural labourers. They work on the field from morning to evening on contract basis, in lieu of grain and cash, either on daily or monthly basis. They are mostly landless labourers.

Other economic activity of the villagers are handloom weaving. There are about 12 persons engaged in this industry out of which about 7 persons are males and 5 females. The Co-operative Society supply raw materials to the weavers with the understanding that the finished products should be handed over to this Co-operative in lieu of cash money in the form of labour charges for the finished products.

About 26 persons were engaged in other works in the town areas of Bargarh or in Hira Cement Factory in Bardol near Bargarh or in Attabira town areas.

The village is well situated on NH 6. The village is well served by Railway lines. Local taxis are also regularly plying from Sambalpur to Bargarh through this village.

All these economic activities clearly show that this village is not economically backward. There are 315 hectares of agricultural land belonging to this village. Major portion of these lands are fertile and have been brought under cultivation. The presence of canals have brought prosperity to these agricultural workers. Two or three crops are generally grown round the year in these lands. Therefore, this village is self-sufficient and even surplus in their production of foodgrains. Economically therefore, this village is well-off than Jamadarpali and Balabaspur village which subsist on marginal economy even though they are situated within the command area of the Hirakud Reservoir.

Conclusion

Economic and social backwardness are characteristics of these villages. Even now when many parts of the country are aiming at economic and social advancement in the context of modern age, these villages remain completely quiet and backward. This may be due to lack of education of the people of these villages. Added to this is the lack of regular exchange of ideas with more progressive areas of the State. There is not a single High School in these villages and none has gone up to High School standard of education. Newspapers do not reach these villages. The villagers do not even listen to Radio Broadcasts. The villagers therefore are ignorant of any programme enforced by the Government for their benefit.

The main occupation of these villagers as we have seen is cultivation. Most of the villagers within the Canal Command Area of the Reservoir get irrigation facilities. But no improved methods of cultivation or modern types of implements are used leading to low productivity, inspite of irrigation. This is mainly due to inferior cropping pattern, low agricultural prices and low per unit value of agricultural production. The other reason for this low production of crops is comparatively low proportion of double cropped areas within and outside the command areas and comparatively low harvest price of rice in the whole State. The other reasons of low productivity is poverty of the rural people of Orissa, leading to low production in the agricultural sector. Line sowing, the Japanese method of paddy cultivation in place of present practice of broadcast sowing, proper crop rotation, deep ploughing, application of better implements and dry farming are to be introduced to get the maximum yield from farm lands of this area. In addition to this, contour bunding, terracing and construction of field bunds are absolutely necessary to prevent erosion of the open and undulating surface of the areas. The production loss due to disease and pests is estimated to be quite substantial every year. This can be reduced to minimum by introducing clear cultivation methods instead of the present system of indigenous seeds. This system can be replaced by introducing improved varieties of seeds with disease resistant strains. It is estimated that only 10 per cent of the existing crop lands, utilize improved varieties of these seeds as compared to 70 per cent of the crop lands utilized in the neighbouring State of Andhra Pradesh.

The soils of these villages are red soil – poor in nitrogen, phosphoric acid, humus and lime. Therefore, they are not very fertile. But these lands respond very well to irrigation, giving good production. But irrigation facilities are provided to only those villages falling within the Command Area of the Hirakud Reservoir, forcing other villagers to depend only on monsoon rain.

Wells are very difficult to construct as they must be very deep to keep water in them throughout the year. But deepening of wells are difficult because of hard rock layers at relatively shallow depths. Even tubewells face the same difficulties. Only big tanks can serve the purpose of the poor villagers to step up production of paddy crop on which the majority of the villages of Orissa depend.

10

CASE STUDIES OF MAHANADI DELTA

Introduction

The Mahanadi River Delta (85°30' to 87° E ' longitude and '19°45' to 20°40' N latitude) comes under the jurisdiction of three administrative subdivisions *i.e.* Kendrapara, Cuttack and Puri comprising 23 Police Stations covering 9000 km^2. The present study with the help of satellite imagery (MSS Bands) is confined to the lower course of the Mahanadi river after the Satkosia gorge. The Mahanadi river is one of the major rivers of the Peninsular India which drains to the Bay of Bengal. All the findings of the MSS Bands are mapped and the boundary is delineated. Earlier it was difficult to delineate the delta but now it could be easily demarcated with the latest technology available. With the help of satellite imagery, landform and landuse maps of the delta have been prepared. Actually the Mahanadi in combination with the Brahmani and the Baitarani have been able to form an extensive alluvial tract extending from Chilka Lake in the south to Balasore in the north, 175 km long and over 75 km wide.

Formation of Delta

The Mahanadi river delta is not the only silt carrying river but two other rivers are also responsible for delta building activities. These two rivers are known as the Baitarani and the Brahmani. Actually three rivers flow parallel to each other in the deltaic area. These three rivers are responsible to form vast alluvial delta of Orissa. Before the construction of Hirakud Dam on the Mahanadi river, the river itself was most erratic in its discharge and every year there were apprehensions of floods during the monsoon season (July to October) . But the recent MSS Bands in false colour combination show the real characteristics of the land at present as the giant Mahanadi river now carries little volume of water which is flowing through the gutter channel in the river plain. The satellite imagery shows the flood zone of the delta. This could not be traced easily earlier by conventional investigation. It is understood from the observation of the imagery that this flood zone perhaps has its genesis due to eustatic change of coast line which sets parallel sand dunes. The eustatic change is clearly deciphered in the coast line. From the collateral information, it is suggested that after the seventh stage of delta formation there was an isostatic change in the coast line which resulted in the emergence of three sets of parallel sand dunes all along the coast, which disturbed considerably the former pattern of drainage. It is clearly visible that the delta has got its maximum growth not in the centre but in the northern part. Prior to the formation of the delta the original shore line was concave towards the sea owing to the hard bed rock of the Ghats. As a matter of fact, the longest distance from the present deltaic apex is supposed to be in the

FIG. 10.1 : Satellite Imagery Coloured Band 5 and 7 (MSS) of Mahanadi Delta (Scale 1:1,000,000)

centre of the convex delta head.

From the satellite imagery the following observation are made (Fig. 10.1). The delta is broadly developed into three broad zones.

(1) Older delta,
(2) Intermediate portion of the delta; and
(3) New and active delta.

But collateral investigation gives the opinion that the delta falls into two broad divisions which are as follows :

(1) The old delta of the older alluvium of the Tertiary period.
(2) The active delta of the newer alluvium of the Post Tertiary period. Of course this collateral information may be accepted but the interpretation has gone beyond this. The Delta mainly consists of Tertiary formation lying adjoining to the old peninsular hillocks marked clearly red colour indicating forests on the hillocks. The adjoining portion of the older delta is mainly of Dharwarian age, consisting of Khondalites and Charnokites.

The intermediate delta is formed of erosional and residual deposits, now being used in farming. The main activity of this zone lies at a higher level than the flood plain proper.

The active delta consists of new alluvium. The flood plain zone is marked by eustatic change of sea level which is of recent origin particularly in Post-Tertiary period, belonging to Pleistocene and Recent age. The advance of sea level was probably demarcated as flood zone. The MSS Bands do not provide clear definition of the older delta due to changing pattern of landscape made by human interference. The scenario is completely changed by the growth of settlements. Most of the settlements are confined in the new delta area along the embankments. The linear stretching in the satellite imagery demarcated as red, i.e., settlements with trees which are of the coastal type of evergreen forest. A little away from the older alluvium, there are isolated forests. It is marked by the FCC (false colour composite), that the laterites are formed in the upper part of hills which are followed by red soil.

Geology

Major portion of the delta area is covered with Recent Alluvium. The surface geology is obscure. Sedimentary sequence ranging in age from Cretaceous to Recent resting on metamorphic basement has been encountered in the Mahanadi delta towards the west and north-west of Cuttack and Bhubaneswar. There are rock exposures of Upper Gondwana, Charnockite, Khondalite granite and gneisses with dolerite dykes and laterites.

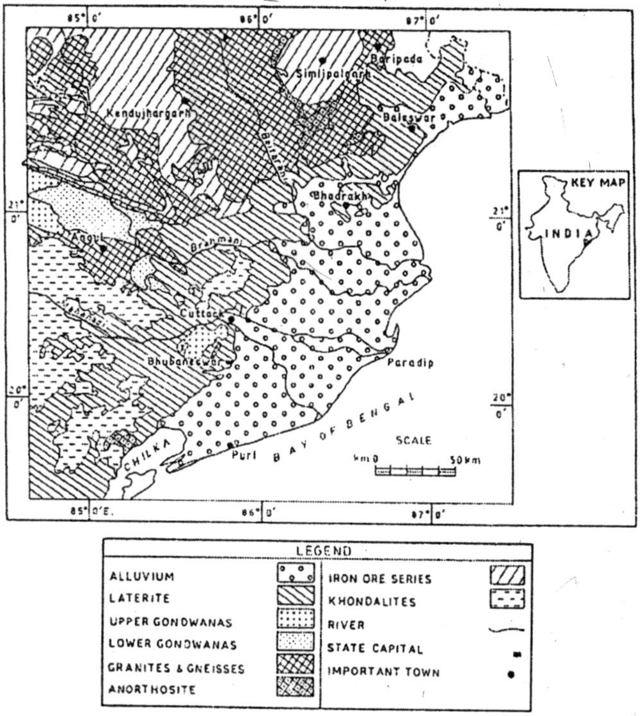

FIG. 10.2 : *Geological Featurs of Mahanadi Delta*

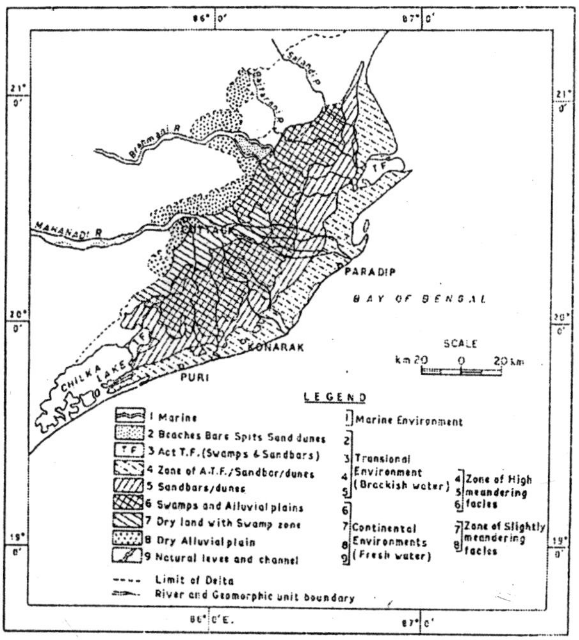

FIG. 10.3 : *Coastal Features of Mahanadi Delta area*

Table 10.1 : Stratigraphy of the Mahanadi delta and its Surroundings

Age		Formation		Lithology
Recent	....			Alluvium
Plio-Pleistocene				Laterites
Mio-Pliocene	...	Baripadabedi	:	Clay and fossiliferous limestone
Lower Triassic to Cretaceous (upper Gondwana)	:	Athgarh sandstone	:	Medium to course sandstone and shale
		Mahadeva series	:	Sandstone and shale
Lower Triassic to Carboniferrous (Lower Gondwana)	:	Damuda series	:	Fine to medium sandstone, shale and coal seams
		Talcher Group	:	Diamictite, fine to medium sand stone, shale and rythmites.
Precambrian and Older Metamorphics	:	Gangpur Singhbhum group	:	Meta-sediments and volcanics.
		Koira-Simlipal group	:	Meta-Sediments and volcanics with iron ore bands.
		Iron Ore Group	:	Quartizite with meta-volcanics and Iron ore.
		Granite Gneisses.		Charnockite and Anorthosite

Source: Quarternary Deltas of India G.S.I. 1991

Towards of the west of the Mahanadi delta, a vast tract of granites, Khondalites, Charnockites and anorthosites of the Eastern Ghats are exposed and form the detached hillocks stretching from NE to SW. The information about iron ore towards North of Eastern Ghat Group include meta-sediments, metabasics and ultrabasic rocks. Lower Gondwanas are exposed towards the southern part of the Iron Ore series. Upper Gondwanas having a general strike from to ENE -WSW are exposed between Cuttack and Bhubaneswar and are comprised of Athgarh sandstone and shales, some of which are ferruginous. The Mio-Pliocene belts of Baripada (NNE of Bhadrak) extend southwards to the alluvium and laterites of Pleistocene to Recent age. The Baripada beds, of Orissa which are of stratified clays and fossiliferous limestones, indicate that during the late Tertiary period, extensive marine transgression took place so that the eastern coast areas of thick alluvium were submerged under the sea.

Geomorphic Aspects of the Delta

The coastal plain of Orissa extends from Chilka lake in the south to Subarnarekha river in the north over a distance 300 km in NE-SW direction. Its width varies from 15-100 km, being in the Mahanadi deltaic region. Relief of the area is very low with slight elevation towards the west. Vegetation in the area consists of tropical evergreen forests. This coastal geomorphic province is drained by the Mahanadi River System namely, Mahanadi, Brahmani and Baitarani. The vast deltaic plains extends for about 170 km from Bhadrak in north to Chilka in south. The Baitarani, Brahmani and Mahanadi rivers form an estuarine type of delta. The extreme southern part is characterised by the presence of a big lagoon, the

LANDFORMS OF MAHANADI DELTA, MSS BAND 4,5 AND 7 FCC

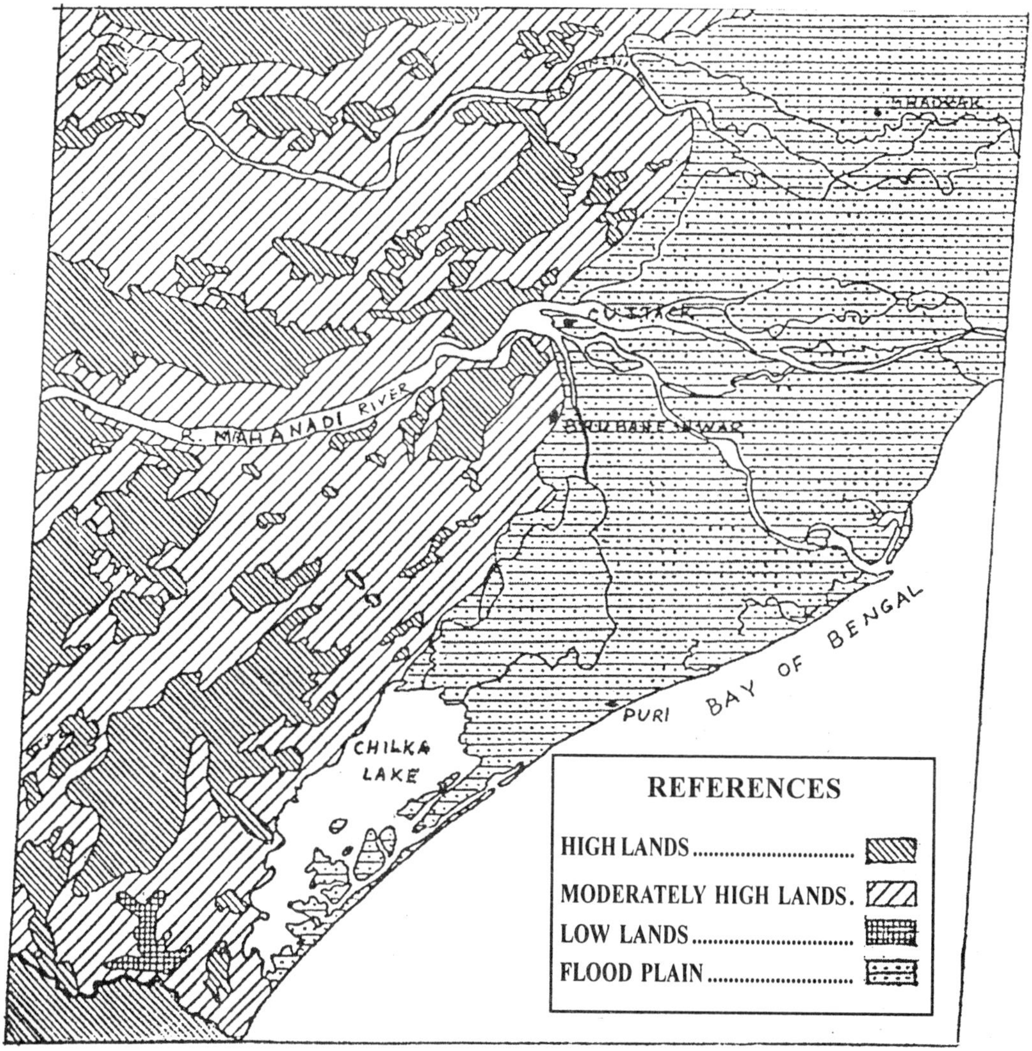

FIG. 10.4 : *Landforms of Mahanadi Delta*

Chilka lake. Mainly three belts of land surfaces are deciphered in the False Colour Composite (FCC), which are as follows :

(1)	High lands	:	75 m to 150 m .
(2)	Moderately High lands	:	20 - 75 m.
(3)	Flood Plains	:	0 - 20 m.

The amount of land covered by low land is very negligible. Only a few patches are identified. The morpho-stratigraphic units are —

(1) Older Deltaic Plain which corresponds to high lands;
(2) Younger Deltaic Plain corresponds to moderate high lands;
(3) Recent Deltaic Plain corresponds to Flood plains. Each unit is characterised by distinct soil type. The alluvial soil predominates, though its age and nature of deposition have brought variations in the soil composition.

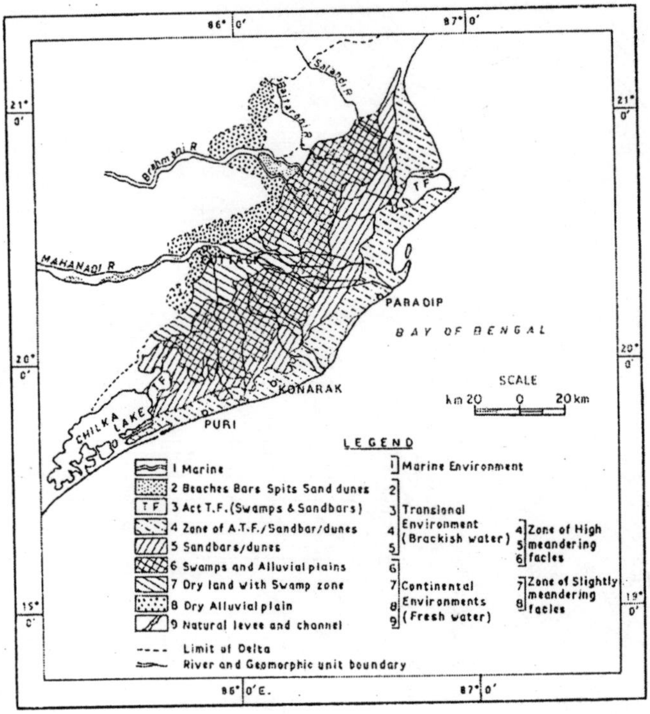

FIG. 10.5 : *Geomorphological features of Mahanadi Delta Area.*

In the Settlement Report, soils of the Mahanadi delta have been classified into nine types which are as follows :

A. Greater percentage of clay

1. Matal (stiff clay)
2. Chikita (strong clay)
3. Chawria (hard clay)

B. Equal percentage of clay and sand

4. Dorasa (mixture of sand and clay in equal proportion)
5. Tel Balia (larger admixture of sand with clay than with Dorasa).
6. Balia Matal (rich sandy loam).

C. Greater percentage of sand (sand group)

7. Balia (sandy)

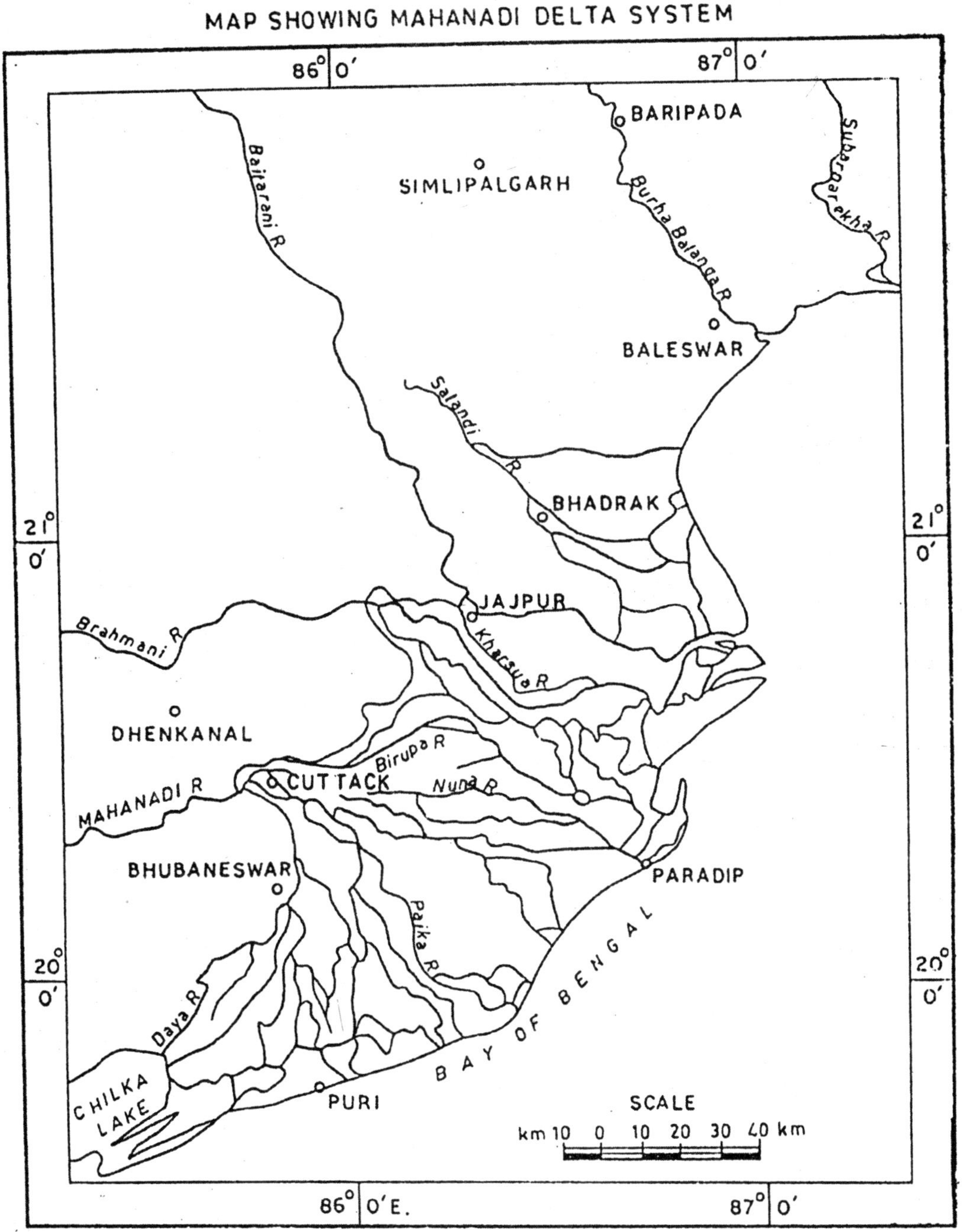

Source : Fig. 13, Page 51

FIG. 10.6 : *Mahanadi Delta System*

D. Loamy soil (Patu or alluvial group)

8. Patu (silty clay)
9. Pankua (sandy with decayed organic matter)

E. Saline soil.

The analysis of the Mahanadi delta soil shows that the soils with higher pH values are met with closer to the coast and in the irrigated region of the delta.

Table 10.2 : Soil Analysis of the Mahanadi Delta

Place		pH value	Loss on ignition	Moisture content
	A horizon	7.0 - 7.5	0.345	0.95
Cuttack	B horizon	7.5 - 8.0	0.315	0.086
	C horizon	6.0 - 6.5	0.305	0.385
Kendrapara		7.0 - 7.5	0.347	0.10
Puri Forest Soil		7.0 - 7.5	0.180	0.003
Balugaon Lagoon Soil		8.0	0.175	0.385

Deltaic Elements

The Mahanadi delta can be divided into three main sedimentary elements, namely continental, transitional and marine. A variety of geomorphic units are recognised in each of these elements. The continental elements include zones of meandering belts and braided river system of upper deltaic plain. Four sub-units viz. natural levee and channel, dry alluvial plain, dry land with swamp zone and alluvial plain of this environment are clearly seen in the Mahanadi delta area. Sediments are predominently sandy and generally micaceous. Fine grained sediments, i.e., silts and clays are deposited in the adjacent fresh water swamps and contain fair amounts of plant remains.

Within the transitional environment the following geomorphic units have been recognised. These are:

— Mangrove swamps
— Marsh
— Flood plain and abandoned tidal flats of the lower deltaic plain
— Beaches, barrier bars, sand dunes/beach ridges
— Tidal flats and lagoons

The marine environment in the modern Mahanadi delta area includes the submarine part of the delta.

Drainage Pattern

The Mahanadi river flows down slope towards the south-east and joins the sea. The river does not follow the basement trend in the strict sense. The Mahanadi river is bifurcated into several channels namely, the Mahanadi, Katjuri, Khowkhai, Devi and Bhargavi near Cuttack. They move radially from one another, which is very conspicuous.

Anastomosing, braided and occasionaly dendritic drainage pattern are generally exhibited in the area. A rectangular type of drainage pattern is observed in the north of the Brahmani river valley away from the coast, possibly characteristic of Iron Ore group terrain. Abrupt and localised braiding, local widening and narrowing of channels and levees and anomolous shapes in the morphological forms of drainage are often observed and may be controlled by neo-tectonic activity. Deviation of the courses of the rivers and its major distributaries, meandering channels corresponding to the cross trend could be related to the presence of Recent geomorphic upliftment of the area.

Deltaic Evolution

Modern Delta

Based on the geomorphic evolution of the coastal plain of West Bengal and Orissa, Niyogi (1975) concluded that delta building processes of the existing set of rivers in the area continued throughout the Quarternary. The generally south-easterly direction of the flow of rivers cut across the Eastern Ghat and Palaeozoic rift trend, indicative of a case of superposition. The Birupa, a distributary channel of the Mahanadi is cut by the Brahmani down stream of Cuttack. Further north-east, the Brahmani in turn is cut by the Baitarani. These suggest that the Mahanadi delta is older than that of the Brahmani, while Baitarani delta is the youngest in the Orissa deltaic plain. It further indicates that the modern Mahanadi delta system in the area is actively building in the northerly direction. Present day progradation of the deltas is indicated by the development of barrier spits in the coastal belt. The river system in the area exhibits different types of deltaic elements. Presence of the abandoned deltaic element clearly suggests more than one major stages of delta formation which were controlled by former strandlines. In the Mahanadi delta, the western limit of the tidal flap corresponds to strandline developed during the last phase of the rise of sea level during Quarterary period. Shifting of river courses have also been indicated by the abandoned river channels. Based on airphoto study, Sambasiva Rao (1978) recognised 3 to 4 strandlines and about 10 meander belts, besides morpho structures and a few lineaments and fractures. The Mahanadi river system shows (i) distributary channel sands associated with inter-distributory silts and muds and point bar sands adjacent to the meandering rivers (ii) tidal channel filter with gravel and clays (iii) barrier bar sand with silty clays. From overall consideration, it is clear that the modern Mahanadi delta system represents a composite progradational delta system which is actively building in a general northern direction.

Ancient Delta

Major deltaic sedimentation in the Mahanadi basin started on a peneplained surface possibly towards the close of Early Middle Miocene period and continuing since then. Collateral information gives the view that due to the prevailing palaeogeographic conditions and various phases of marine transgression and regression, considerable variation in the thickness of the sedimentary rock is observed specially in the off shore wells drilled within the various depressions. The presence of fairly large thickness of Miocene deltaic sediments is observed in these two major depressions. Sediments deposited in the flood plain of this lower Miocene delta comprises of interbedded clays, silts and fine grained sands with occasional lignite. Coarser sands with occasional clays are deposited in the river channels. The shelf edge during the lower part of Middle Miocene was located in the northwestern part of Mahanadi off shore basin. During the rest of the Middle Miocene to late Miocene, there is an increase of influx due to which, there has been a slow rate of sedimentation. The continental shelf was much wider during the Lower Miocene period than in the Middle and Upper Miocene. Deltaic activity was pronounced during the Mid to Late Miocene period. Progradation of the shelf margin during the Mid Miocene to Recent period shows successive movement towards the sea. The delta front also progressively moved further towards south-east during this time. Shifting of the Middle Pliocene shelf margin is minimal in the central area. From Mid-Pliocene onwards, the sedimentary influx had substantially increased in the central part. Varied deltaic sediments in terms of elements and extent of different periods are therefore preserved in the offshore areas related to the ancient Mahanadi delta system. During Late Palaeocene-Eocene era intensive deltaic sedimentation took place in the coastal *Puri-Konark depression* where as transgression of the sea in the subsiding Chandbali depression led to the development of marine sediments. In the Cuttack depression however fluvial condition prevailed till Recent age. During Miocene, intensive deltaic sediments were deposited in the Paradip depression and *Puri-Konark depression* and adjoining, offshore areas, whereas, transgression of sea continued in the Chandbali depression. Deltaic sedimentation continued in the *Puri-Konark depression* till the end of Plio-Pleistocene whereas in the Paradip depression it continued till the end of Late Miocene when fluvial to estuarine environmental conditions were established.

Population and Settlement

Mahanadi delta system is the home of 9 million people and has got an average density of 420 persons per km^2 (1991). Thus, although it accounts for 6 per cent of the total area of Orissa, it accommodates 15 per cent of her population. The population density figure ranges from 215 to 306 per km^2 during (1931-61 and from 306 to 420 per km^2 during 1961-91). While it is the generalised picture of the delta as a whole, the density figures range from 3394 persons km^2 in case of Puri municipality to as low as 201 in Brahmagiri Police Station of Puri District. It has also been observed that the density of population is higher on the Northwestern side in Cuttack district than in Puri district which is southernly situated. The density of Cuttack district is 459 persons per sq.km. and in Puri, it is 352 persons per sq.km. in 1991. This high density is due to (1) irrigation (2) transport and (3) good quality of fertile land, etc. So it is observed that the density increases from the delta mouth to the apex and from the non-irrigated and unprotected to the irrigated and protected tracks of the delta. For example Govindpur, Balikuda, Mahanga and Salipur have high density (over 500 persons per sq.km.), being located in the flood protected and irrigated tracts, whereas, Patamundai, Gop, Mahakalpura, Rajanagar, etc. have low density being located in the lower reaches. Bhubaneshwar, Puri and Cuttack are the prominent cities in the delta. Bhubaneshwar, is the new Capital of Orissa and in the recent past the capital is shifted from Cuttack which has a commercial and industrial base. Puri is famous for its Lord Jagannath's Temple and has religious-*cum*-administrative outlook. These three urban centres constitute 63 per cent of the total urban population of the state according to 1981 Census. Except these three urban centres and Puri, the delta is dominantly rural in character with about 90 per cent of the total population living in the country side. In most of the coastal area linear pattern of rural settlement is observed and in the satellite imagery (FCC), it is clearly observed by stretching of red filament which shows that most of the settlement is covered with trees and these settlements are mostly situated on the higher land, that is on the levees. These settlements may be called as 'levee settlements.' 'In these levee settlements both permanent and temporary habitations are found. Temporary huts are located near the coastal margins in the areas which are liable to inundation by high tide water. The houses cling to the roads or high levees and avoid river sides which are not free from the danger of floods. As the general level of the land is low, settlements have been established over high levels and along embankments which provide safety from floods. Most of the temporary huts are fishermen's settlement. From

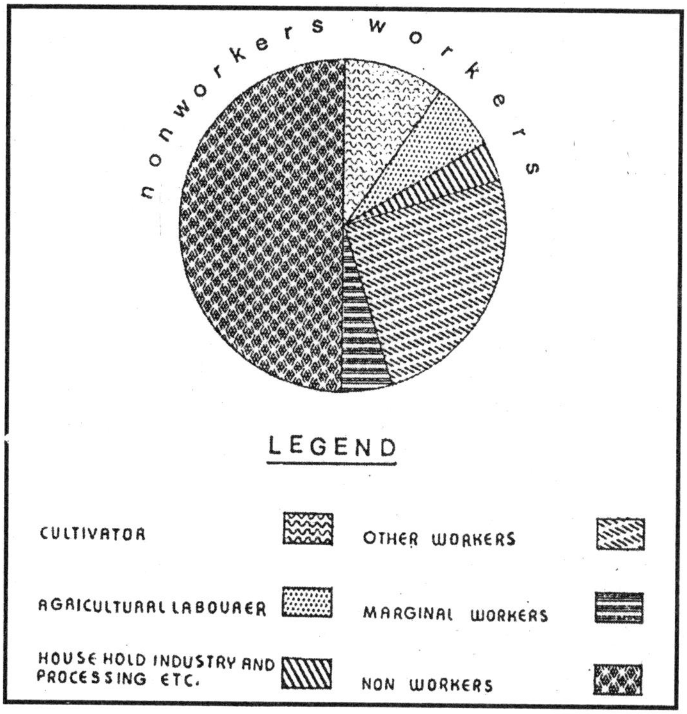

FIG. 10.7 : *Occupational Structure of Mahanadi Delta*

the agricultural point of view the more important tracts which are protected from floods and have irrigation facilities, the density of settlements is fairly high, settlements are of large size and are often located along the canals and distributeries. The occupational structure of the Mahanadi delta is greatly influenced by the agricultural economy. Occupational pattern is not diversified compared to Orissa as a whole as two-thirds of the people depend on agriculture in the delta. Small household industries account for 2.96 per cent, which are also primarily agrobased. In transport storage and communication about 1.14 per cent the workers are engaged. About one-fifth of the total population depend on various services. According to 1981 census 10.25 per cent of the working population is engaged in cultivation and 6 per cent as agricultural labourer. Whereas in Household industry, processing, servicing and repairing only 3.1 per cent of workers are engaged.

Table 10.3 : Occupational Structure of Mahanadi, 1981

		(in p.c. to total workers)
I.	Cultivator	10.25
II.	Agricultural labourer	6.3
III.	Household industry, manufacturing, processing, etc.	3.1
IV.	Other workers	26
V.	Marginal workers	5
VI.	Non-workers	50

Agriculture

The cropping pattern is the end product of the complex factors that are operating on the land like

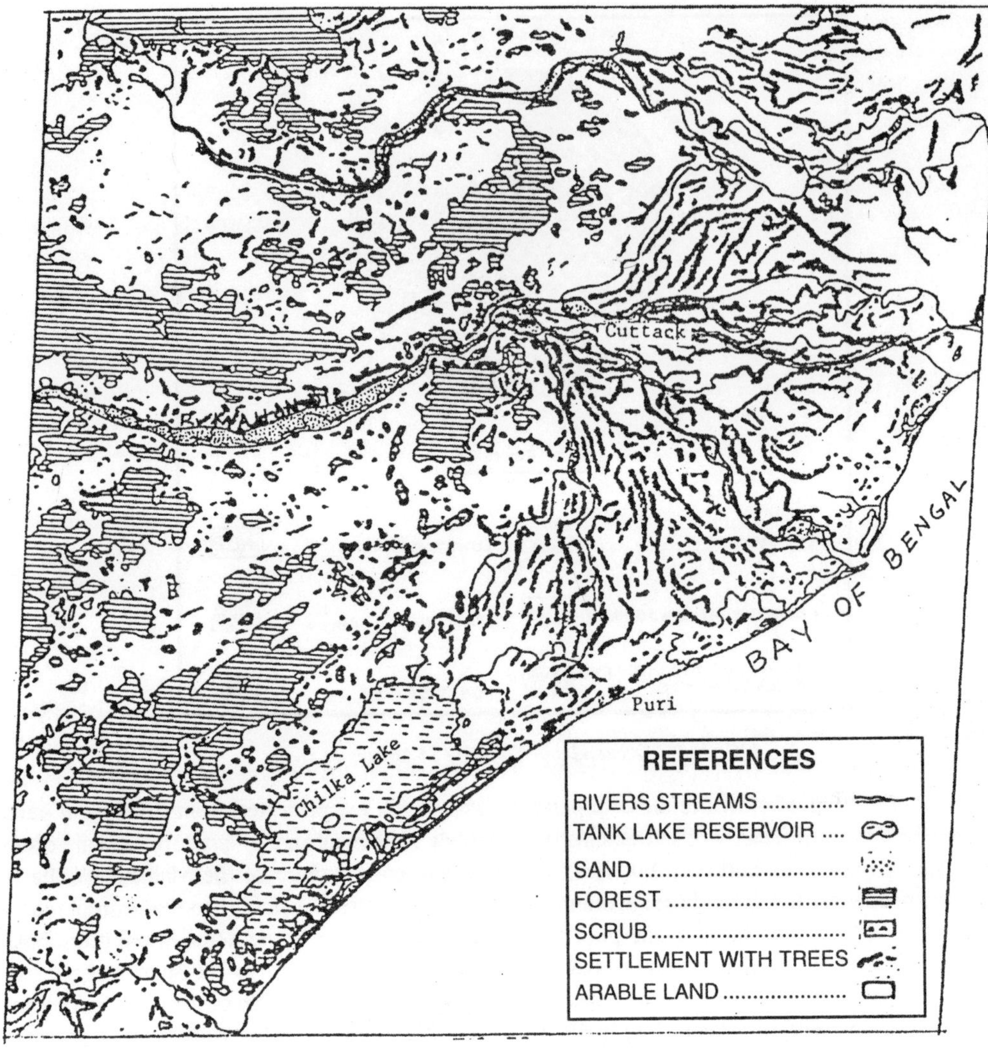

FIG. 10.8 : *Landuse of Mahanadi Delta, based on MSS Land 4,5 and 7 FCC*

terrain, climate, soil, availability of fresh water and availability of worker. In the Mahanadi delta 45 per cent of the area is under cultivation and this figure is due to the presence of a large number of water bodies, marshy swamps of different sizes. Considerable area is occupied by the distributaries of the Mahanadi, creeks, canals. The gross area under cultivation is supposed to be 70 per cent of the total area of the delta. Due to extension of irrigation facilities as much as 35 per cent of the gross cropped area is double cropped. Canal irrigation scheme was undertaken in 1863-64 by a private company under the guidance of Sir Arthur Cotton and water for irrigation was first supplied in 1869. The canals are categoried as perennial and inundation canals. Prior to the construction of Hirakud Dam, there are three anicuts on the Mahanadi river system. These are anicuts are at (i) Naraj at the head of Kathjuri which serves as feeder anicut by diverting water to the Mahanadi (ii) at Jobra on the Mahanadi (iii) on the Birupa which helps in making an artificial lake at the head of the confluence of the Mahanadi. The water released through the power houses of Burla and Chiplima, irrigate further 436000 hac. of canal command are in the Mahanadi delta in the district of Cuttack and Puri, in addition to stabilizing irrigation to 78540 hac. in the same district.

Delta Irrigation System

The water discharge through sluices at Hirakud Dam records an increase by nine times. With the availability of water, new areas are brought under Canal Command Area and the irrigation capacity of the Mahanadi division is increased. Another weir is constructed across the Mahanadi, further up-stream from Naraj and the main canal is taking water from the right bank. It has to cross the Kharkhai by a Syphoon and then enter the Devi-Prachi Sector. In all there are eight main canals. All the major canals are contour canals for which flow irrigation is possible. They have got to irrigate 430860 hac. including 216310 hac. in the formerly irrigated region.

Minor Irrigation

Minor irrigation is of greater importance in the area where canal irrigation is not available. The net addition is only 280 hac. to a total of 214, 560 hac. of irrigated land of delta. The lift irrigation from the wells and tanks are practised as the water table is not far below the surface.

Cropping Pattern

Cereals dominate the cultivated crop this group accounts, for 77 per cent of the gross cropped area. The cash crops are jute, sunhemp, tobacco and sugarcane. This group accounts for 5.25 per cent of the gross cultivated land. Amongst the cereals, rice dominates and three types of paddy are being cultivated. The Sarada or winter paddy is best suited for low lands subjected to flooding. In irrigated tracts also it grows well. The Biali (Autumn paddy) thrives best in the Sarada lands which are slightly elevated and free from frequent floodings. The *Dalua* (Summer paddy) grows well in the marshy lands reclaimed by suitable drainage methods in the areas subjected to constant flooding. It is harvested before the on-set of monsoon. Such areas are found in lowlying lagoon areas of Samagara Patmar Puri. Among the cultivated crops rice dominates as it accounts for 77 per cent of the gross cropped lands and thus, makes it primarily a mono-cultural region. The percentage of rice cultivation varies from 97 per cent to 44.5 per cent. The highest percentage of rice cultivation is noticed in Brahmagiri and in Mahanga. Jute crop is most important cash crop in the delta and it account for 5 per cent of the gross cropped area of the delta.

11

CONCLUSIONS

The science policy of the Government of India is guided by a resolution adopted in Parliament in 1958. One of the main objectives of the resolution is to secure for the people of the country the benefit of acquisition of scientific knowledge and its application. It is also said that the knowledge acquired by the scientific communities should not be confined amongst themselves. The knowledge should be disseminated to the general masses in a simple way so that they could understand its implication. In the context of this dissertation, application of Remote Sensing for the benefit of society has multiple functions namely in (1) Landuse, (2) Landform, (3) Cartography, (4) Geology, (5) Hydrology, (6) Crop Production and its protection from natural hazards like cyclone, insects, animals, etc. (7) Soil Resources to protect soil erosion and effective measures for soil protection. (8) Forestry , Aerial photographs and satellite imagery can monitor the growth and depletion of forest in the interior and remote areas.

In this book remote sensing techniques are applied in a part of the Western Himalayas and most of the studies are confined in Dehra Dun and Rishikesh area. So far the satellite imagery has identified four major sets of tectonic lineaments and several skewed alluvial fans. The lineaments possibly reflect presence of upper crustal structure that have largely experienced extension as a consequence of continental collision. Evidence for possible morphotectonic activity controlled by till block system can be obtained from isolated accumulation of the Quarternary sediments within the assymetrical tectonic basin in the Himalaya in Dehra Dun and Rishikesh. Academicians know very well the theories of Remote Sensing/Geographical Information System, but they lack proper instrumentation and application. It is the ardent duty of the scientists to bring the applications of Remote Sensing to common people, so that they could understand the new dimension of science.

During the construction of Hirakud Dam, the local inhabitants apprehended that such water would damage the land. But people from distant places and adjoining State of Andhra Pradesh are effectively utilising the irrigation water and getting its benefits. These migrants have settled on both sides of NH 6 at Gurbhaga and Attabira. The cultivation scenario has completely changed in the Hirakud Canal Command Area, after the completion of the project, which is now producing rice three times in a year.

To study landuse with the help of aerial photographs and satellite imagery, it is essential to get a comprehensive idea of the area to be investigated. Hirakud Canal Command and its adjoining areas have been taken as an example in the context of the dissertation. With the completion of the project, agriculture has recorded spectacular growth with assured irrigation. Many rural settements have emerged in the region and some others have disappeared. In this context it has become necessary to take suitable measures to adopt a rational orientation of landuse in the area. With this idea in view detailed survey of six villages and three towns have been carried out to identify the present pattern of landuse, so that plans for the future may be chalked out, integrating rural and urban requirements. It is thus evident that with the modern technology of Remote Sensing, planning for regional development could be taken out over a large region in limited time. Recently, the Department of Science & Technology, Government of India, has taken up a project to prepare maps of all the districts separately which is known as District Planning Map Series with the help of Thematic Mapper (TM)/IRS LISS II blown up in the scale of 1:250,000. This would provide the Landuse and Crop Study in greater detail with reasonable degree of scientific accuracy.

Morphology, drainage, geology and soils of selected regions, which are partly accessible and partly inaccessible have been deciphered with this technique. The two selected areas are located in the foot-hills of the Himalaya. These regions are Dehra Dun and Rishikesh. At Rishikesh, after completion of air photos and satellite imagery interpretation, field traverses are carried out to correlate and establish soil boundaries with the photo-interpretation lines. Initially the sampling are traversed. Part of the sampling area is covered, where there is almost absence of reference points needed for orientation and location of auger boring and pit sites. The sampling areas are selected in such a way that it includes all land types which are not easily accessible. Moreover, the aerial photographs are too old to show the present picture of the terrain. On the basis of such air -photographs much time is spent to locate the observation sites and orientation with the help of compass. Further there is much variation of the photo scale resulting in wrong location, in spite of the use of compass. Frequent rectification of the same reduced the efficacy and capacity of the work. In order to overcome these difficulties in the field, emphasis should be given in the following :

(1) The photograph should be recent,
(2) Photo scale should not vary greatly, i.e. the photograph must be vertical as far as possible without appreciable variations in flying heights,
(3) Sample area should be selected within easy reach and it should represent all the physical and cultural elements, and
(4) Before field mapping of the sample area, a few rapid reconnaisance trips should be carried out in order to acquaint oneself with the area. Once the area is known, photographs will help in drawing neat thematic maps.

While drawing final maps, special care has been taken to correlate soil boundaries with the photo interpretation lines based on field observation of augur boring and morphological description of the profile pit. The physiographic boundaries are correlated with the soil boundaries. The interpretation in respect of vegetation pattern is often correlated with the soil pattern. However, in some cases such as on the oldest terrace of the Chandna Rao, the correlation could not be worked out in spite of non-distinctness of vegetation pattern. Conventional methods of soil survey had to be carried out to establish the soil boundaries. In the *Terai* the prominent terrace bluffs could not be located on aerial photographs because of poor visibility at the time of the photography. Such handicaps can however be successfully overcome by satellite imagery. The aerial photographs are now used as base map and to get the synoptic coverage the Satellite Imageries are used. The Satellite Imagery clearly shows the plain land, the Siwaliks and intervening Pleistocene terraces, followed by Central massive and permanent snow covered area.

In short it may be pointed out that modern scientific mapping has been much benefitted with the development of Remote Sensing. The satellite Imageries belong to different bands. Some of them have greater depth which not only represent the surface features but also record below them. For example in Band 5 the terrain is shown in considerable detail. Band 7 on the other hand represents the water bodies as well as the continental shelfs in reasonable degree of accuracy.

Satellite Imageries give perspective ideas of the terrain pattern, drainage, soil cover, landuse pattern, settlements over a region for a particular period. In field survey it requires considerable time and man power to cover the same area. Moreover, in case of field survey the relevant information suffer from lack of standard, because of the association of numerous personnel whose ideas are quite obviously different from each other. In Satellite Imageries, the observation are standardised and comparable. As such mapping based on satellite Imageries is reasonably accurate, specially when large areas are mapped with limited manpower and resources.

The aerial photographs, on the other hand, give detailed information. But to cover large areas large number of air-photographs need to be consulted. The photographs become hazy if weather is bad. In such cases, fresh photographs are necessary. Moreover, photographs over wide areas cannot be taken within a day. It may take months when the pictures may change from the initial ones.

To get the best result, be it in case of terrain analysis, land use, resource studies or preparing plans for integrated development of any country, it is necessary to take the help of Satellite Imageries and aerial photographs. For scientiffic accuracy the geographic information available from the two techniques should also be checked and cross checked by detailed field investigations.

The researcher has used these three techniques in his investigations outlined in the present book. Mappings have also been made with the help of Satellite Imagery of 79 m and 30 m resolution. Further details have been gathered from aerial photographs. A number of field checkings have also been carried out in the areas of investigations.

BIBLIOGRAPHY

Abdulgani, S. A. (1963) : *Soil Survey of the Island of Texel,* Thesis I.T.C. Delft.

Abram, T. (1944) : *Essential of Aerial Surveying and Photo-interpretation,* New York.

American Society of Photogrammetry (1960) : *Manual of Photographic Interpretation,* Washington D.C.

American Society of Photogrammetry (1965) : Photo-interpretation in Space Science, *Photogrammetric Engineering,* Vol-31, 1065-75.

American Society of Photogrammetry (1983) : *Manual of Remote Sensing,* Vol-I-1983.

Andahl, A.R. (1958) : Soil Survey Interpretation — Theory and Purpose, *Soil Sci. Soc. of America Proc* 22,, 152-54.

Anderson, J. R. E. E. and J. T. Road (1972) : A Landuse Classification System for use with Remote Sensing Data, *U.S. Geological Survey, Circular* 671, pp. 1-16.

Andronikov, V.L. (1957) : The deciphering from aerial photographs of the Soil cover of Wooded Steppe, *Pochvedenis* 5-70-76 (Russian).

Archives Internationals de Photogrammetry (1962) : *Transactions of the Symposium on Photo-inter pretation,* Delft, the Netherland, Waltman, Delft 1962.

Archieves International de Photogrammetry (1962) : *Symposium Delfts* (Vol.14), Netherlands.

Arens, R.F. Kramer and Ch Langner (1958) : *Uebe die Bodenstruktur,* dan Wasserhanshalt und.

Arnold, A. (1978) : *The Role of Science and Technology in Development Proceeding on Remote Sensing of Environment,* Vol. 1, pp. 175-78.

Atkinson, J. R. and N. N. Brown (1962) : Airphoto-interpretation of Road of South Rodesian Government.

Aubert, G. and P. Duchaufour (1956) : Project de Classification des Sols, *Trans. VI, Int. Soil Sci. Congr.* E, 597-604.

Auden, J.B. (1935) : Traverses in the Himalaya, *Rec. Geol. Surv. of India* 69 (2) pp-123-167.

Baldwin, M.H.M. Smith, White Lock H.W. (1947) : Belder D.J. (1943) : The use of serial photographs in Soil Mapping. *Photographic Eng.* 13, 532-36.

Banerjee, R. K. (1968) : Geomorphic Analysis Based on Aerial Photo Interpretation of Rishikesh area in Ind. *21st I.G.C.* Selected papers, Vol-4, published in 1972.

Banerjee, R. K. (1981) : Application of Aerial Photo-interpretation and Remote Sensing Technique for the Preparation of *National Atlas of India,* I.C.A.

Banerjee, R. K. (1986) : *Simposio Latino-Americano De Sensoria-Mento Remoto,* Brazil.

Banerjee, R. K. (1975) : Hirakud Power and its impact on the Production Economy of Orissa, *Industrial Situation,* Vol-IV, No. 2.

Banerjee, R. K. (1986) : Landuse of Jamdarpali and Gurbhaga village, National Seminar on Landuse Mapping, NATMO, Govt. of India, DST, Calcutta.

Banerjee, R.K. (1986) : Growth of Burla and Hirakud towns in Orissa, jointly with Prof. Maya Banerjee, *Geographical Review of India,* Vol. 48, No. 4.

Banerjee, R. K. (1987) : Study of Levee Settlement System of Nadia District of West Bengal with the help of Aerial Photographs and ' Satellite Imagery and its Mapping, National Seminar on Mapping of Settlement System 1987, NATMO, Calcutta.

Banerjee, R. K. and Mazumdar, K . C . (1967) : Soil Survey Report, I.P.I., Dehradun.

Banerjee, R.K. and Nag, P. (1992) : *Thematic Cartography and Remote Sensing,* Concept Publishing Company, New Delhi.

Baren, F.A. and H. Kiel (1958) : Relationship between the Minerological Composition and Fertility of Tropicai Soils, *Trans. VIth Int. Congress Soil Science.*

Barret, E.C. and Curtis, L.F. (1976) : *Introduction to Environmental Remote Sensing,* John Wiley, London.

Belcher, D. J. (1945) : The Engineering Significance of Soil Patterns, *Photogr. Eng.* II, 115-48.

Belcher, D. J. (1948) (a) : Field Investigation, Aerial Photographic Analysis, *Proceed. II Int. Conf. Soil Mechan. Faund. Eng.* I, 313-21.

Belcher, D.J. (1948) (b) : Determination of Soil Condition from Aerial Photographs, *Photographic Eng.* 14, 482-88.

Bennett, H. H. (1939) : *Soil Conservation,* McGraw-Hill, New York.

Boer, Th. A. de. (1959) : Een Globale Grassland Vegetatie Karte Van Netherland, *V.L.O.* 635, 1956.

Boer, Th. deen and Th. J. Ferrari (1956) : Bodem Vruchthaarhid, Vegetatis Karteringeenheid en Opbrengst van grassland in Eenzandbebied (de Gelderse Vallei), *V.L.O.* 62.15.

Boer, Th. A. de D.V. Diepen and Th. J. Ferrari and APA Vink (1956) : Bodemkartering en Bodem Waarderring in West-Duitsland, Landbauwk. *Tijdschr.* 68, 855-62.

Bomberger, E. Hand, H.W. Dill (1960) : Photo-interpretation in agriculture, Ch. II. *Man. Photogr. Int.* 561-632.

Brink, A. B.A. (1962) : Airphoto-interpretation applied to soil engineering mapping in South Africa, Arch. Int. *Photogrammetria.* 14, 498-506.

Brock, G.C. (1960) : The quality photographic image, *Photogrammetria* 16, 132-158.

Buckhannan, W.H. (1939) : Technique and use of aerial photographs for soil mapping and reproduction of field maps, *Sol. Sci. Soc. Amer. Proceed.* 4.

Burger, A. (1957) : *Photographic Aeriannes et.* a Management *du* Territoire 128 pp. Paris.

Buringh , P. (1954) : The Analysis and Interpretation of Aerial Photographs in Soil Survey and Classification, *Neth. Jour. Agr. Sci. Soil,* 16-26.

Buringh, P. (1961) : Soil and Land Classification in Dutch New Guinea, *Trans. VIth Int. Cong. Soil Science.*

Buringh, P. and W.J. Van Liere (1961) : Example of a Reconnaisance Soil Map Produced by the Pedological Analysis, followed by the study of soil in the field, Vol. 55.

Buringh, P. and C. H. Edelman (1955) : Some Remains about the soils of the alluvial plain of Iraq, South of Bagdad, *Neth. Journal Agri. Sci.* 3, 40-49.

Buringh, P. and K.T. Louay (1961) : Soil Survey and Classification in Iraq, *Trans VIth Int. Cong. Soil. Sci.* : V-14.

Buringh, P. (1960) : The application of Aerial Photograps in Soil Survey, *Manual of Photo-interpretation,* p . 633-66.

Buringh, P., (1960) : The Development of Aerial Photo-Interpretation in Soil Section of I.T.C. for Aerial Survey Delft, Netherland No. H-2.

Bushnell, T. M. (1932) : A new technique in soil mapping, *American Soil Survey Bult.* 13, 74-81.

California Forest Station (1958) : *Soil Vegetation Survey.*

Campbell, J. H. (1981) : Spatial Correlation effect upon Accuracy of Supervised Classification of Land cover, Photogrammetric Engineering and Remote Sensing.

Casagrande, A. (1949) : Classification and identification of soils, *Trans. Amer. Soc. Civ. Eng.* pp. 901-991.

Census Atlas of India (1961) : R.G.I., New Delhi.

Chevrel, M. M., Courtois and G. Well (1981). The Spot Satellite Remote Sensing Mission, *Photogrammetric Eng. and Remote Sensing,* Vol. 47, No. 8; pp. 1163-71.

Chin Tung Kee (1962) : A Report on the Application of Airphoto-interpretation in the Federation of Malaya, *Arch. Int. Photogrammetry* 14, 492-497.

Christian, C.S. (1952) : Region Land Survey, *Journ. Austn. Inst. Agri. Scie.,* 18, 140-46.

Churchiil, E. D. and R. L. Sitt. (1955) : Association Analyses Applied to Interpretation of Aerial Photographs, *Photogr. Eng.* 21, 578-602.

Clarke, G.E. (1957) : *The Study of the Soil in the field,* 4th Edn. Oxford.

Clark, J. (1980) : The Effect of Resolution in Simulated Satellite Imagery on Spectral characteristic and computer assisted landuse classification, Jet Propulsion Laboratory, Pub. 715-22, Pasaden, CA-125 pp.

Col. Vocoressess, A. P. (1979) : Effective Resolution Element (ERE) of Remote Sensors. Memorandum for Record, Feb 8, United States Department of Interior, Geologic Survey, Reston Virginia.

Colwell, R.N. (1956) : Determining the Prevalance of certain cereal crop by means of Aerial Photography, *Hilgardia,* Vol. 26, No. 5, pp. 233-86.

Colwell, R. N. (1975) : The multiconcept as applied on the acquisition and analysis of remote sensing data R.G. Reeves (ed). *Am Soc. of Photogrammetry, Fallschurch,* Virginia, p. 5-11.

Colwell, R.N. (1978) : Keynote address, History and Future of Remote Sensing Technology and Education Conference on Remote Sensing Education, A Workshop held in Stanford University, NASA, American Research Centre.

Colwell, R.N. (1979) : Remote Sensing of Natural Resources, Retrospect and Prospect. Proceeding of Remote Sensing for Natural Resources, an international view of problem, promises and accomplishment, University of Idaho, Moscow, pp. 48-68.

Conklin, H. E. (1957) : *The Cornell System of Economic Land classification.*

Cutrona, L. J. N. , E. N. Leith Procells and W . E. Vivian (1966) : Paper on application of coherent optical processing technique to synthetic aperture radar, *Proc. of I.E.E.E.* Vol. 54, No. 8, pp. 1026-32.

Curtis, L. E . (1962) : *Soil classification and photo-interpretation.*

Custer, S.A. and S. R. Mayer (1955) : A comparative analysis of curricula and Techniques used in training of photographic interpreters, Boston Univ. Tech. No. 19.

Dan, J. (1962) : The disintegration of Nari lime crust in relation to relief, soil and vegetation, *Arch. Int. Photogrammetry* 14, 189-94.

De Haas W . (1962) : The areal photograph as sociological tool ITC.

De Meester, J. (1961) : The application of airphoto analysis in a detailed soil survey. Tanganyika ITC Publ. B/0 Delft, Netherland.

. De Molyn, J . C . (1961) : (a) Geomorphological air photo-interpretation in applying to sand and gravel surveys (Dn E. De Ingenoeur, B 113-B 120).

De Molyn, J. C. (1961) : (b) Air photo-interpretation as applied to civil engineering projects (Dn. E) *De Ingenieur* 73B-123-(B-126).

Dill, H.W. (1955) : Photo-interpretation in flood control appraisal, *Photogr.* Eng. 31.

District Census Handbook, Sambalpur 1961.

District Census Handbook, Sambalpur 1951.

District Census Handbook, Sambalpur, Part B-1971.

District Census Handbook, Sambalpur, Part X-B, Town and Village wise Primary Census Abstract, 1971. & 1981.

District Statistical Handbook, Sambalpur, 1969 and 1970, (combined issue).

Dudly Stamp, L.C.S. (1956) : Report of the Commission on invention of world landuse (Int. Geog. Cong.), New York.

Dubois, P. and W. D. Bruce (1978) : Remote Sensing technology transfer, the Canadian Peruvian of the Twelfth Int. *Symposium of Environment,* Vol. 1, pp. 203-11.

Edelman, C . H. (1949). Land classificatie, Bodemkundige Voordrachten (reeks land bouw No. 9), 124-36.

Edelman, C. H. (1959) : Application of soil survey in land development in Europe, F.A.O. meeting, Oxford, Sept. 1989.

Einevoll, O. (1957) : Photo-interpretation in Soil Mapping and Land Classification, *Norsk Tiddskrift fir Jordskifte* og, *Land maling,* 4, 222-35 (Norwagian Engl. Summ.).

Engene, S.A. (1951) : Getting research data by conference, *Journal of Farm. Economics* (U.S.A.), 33, 847-56.

Estes, J.E. (1982) . U.S. Remote Sensing Policy : A perspective; Proceedings of International Geoscience and Remote Sensing Symposium, Munich, Germany.

Estes, J. E. , J. R. Jeusen and D. S. Simonett (1980) : Impacts of remote sensing on U.S. geography. *Remote Sensing of Environment,* Vol. 10, 72 pp.

Ferrari Th. J. (1950) : Growth factors and soil productivity Trans. 4th Int. Congress of Soil Science, *Armsterderm,* Vol. 1, 348-52.

Ferrari, Th. J., H. Pijl, J.T. Venekamp (1957) : Factor Analysis in Agricultural Research, Netherlands, *Journ. Agri. Science,* 5, 211-21.

Finkel, H.J. (1961) . The movement of barchan dunes measured by aerial photogrammetry, *Photogr. Eng.* 27, 439-44.

Fischer, W.A. and C. J. Robinove (1968) : A rationale for a general purpose earth resource observation satellite, *Proceeding of Remote Sensing Satellite Symposium,* University of Washington.

Fitho, P. Armando Pacheo dos Santos and others (1980) : The use of landsat data for evaluation and characterization of deforested patureland and reforested areas in Brazil; Proceeding of the fourteenth Int. Symposium on Remote Sensing of Environment, *San Jose, Costa Rica,* Vol. III, pp. 1723-29.

Fisk, H. N. (1944) : Geological investigation of the alluvial valley of the lower Mississippi River, Miss. River, Comm. Vickburry, Miss. 78 pp.

Forshaw, M.R., A. Haskell, P.F. Miller, D.J. Stanley and J.R. Townshend (1980) . A Review paper, Spatial Resolution of Remote Sensed Imagery; paper submitted to U.S. Committee for Peaceful uses of outer space, pp. 53.

Francica, J . R. , R . W . Bernie and G . D. Johnson (1980) : Geologic mapping of the Ladakh Himalaya by Computer processing of Landsat data; Proceedings of the 14th Int. Symp. on Remote Sensing of Environment, *San Jose, Costa Rica,* Vol. II, pp. 773-82.

Frankart., R., A. Herbillon and Others (1962) : Bugesera. Mayaga. Busoni, Carte P. Dologique, Rapport preliminaire, report UCLINEAC-ITC Delft.

Frost, R.E. and K.B. Wood (1950) : Air Photo Patterns Soils in the Western United States, Perdue University, U.S. Dept. of Commerce, Techn. Divn. Report No. 85.

Frost, R. E. (1952) : (a) The air photo-interpretation programme of research and instruction at Perdue University, *Photogr. Eng.* 18, 701-19.

Frost, R. E. (1952) : (b), Interpretation of Permafrost feature from air photos, Highway Res. Bd. Special Report 2, 223-46.

Frost, R. E. (1953) : Factor limiting the use of aerial photographs for analysis of soil and terrain, *Photogr. Eng.* 19, 427-30.

Frost, R.E. (1960) : Reading and interpretation of aerial photographs, *Tijdschr. Kon. Ned. Aardr. Gen.*, 65, 452-62.

Gansser, A. (1969). *Geology of Himalaya* : John Willy and Sons

Gaveman, A. V. and Yu Liverovsky (1953). Aerial photography in soil mapping, *Poohvovedina* 3, 1-9 (translated) .

Godrick, F. E. (1981) : The use of landsat data in an operation of forest resource information system (FRIS), Perdue University, W. Lafayette, Indiana, U.S.A. , pp. 543-44.

Goldman, L.M. (1960) : *Application of Colour Photography for the Study of the Terrain,* Moscov, Publication House Geodat, translated.

Goodale, E. R. (1953) : An equation for approximately the vertical Exaggeration of a stereoscope view, *Photogrammetric Engineering,* 19, 607-16.

Goodman, M. Smith (1959) : A technique for the identification of an serial photographs, *Photogrammetric Eng.* 25, pp. 131-37.

Goosen, D. (1961) : *A study of Geomorphology and Soils in the Middle Magdelana Valleys,* Columbia, I.T.C., Pub., &, Delft Netherlands.

Gordon, F . (1980) : The space time relationship of Data points (Pixel) of the thematic mappers and the Multispectral Scanner, NASA Technical paper, 1715, Goddard Space Flight Centre, Greenbelt, Maryland, U.S.A.

Gorrie, R. H. (1953) : Aerial photography as an aid to planning improved landuse and flood control, *World Crops* 5, 405-08.

Gouffon, R. K. (1955) : Gully erosion surveying with aerial photographs, *J. Soil water Conservation* 9, pp. 173-74.

Gould, P. (1981) : Lettering the data speak for themselves, *Anals Associate of American Geographers,* Vol. 71, pp. 168-76.

Guerrero, E. Bata, R. Punongbayan, O. Mendoza (1980) : Geologic and mineral potential mapping in the Philippines using landsat imageries, *Proceeding of the 14th International Symposium on Remote Sensing of Environment*, San Jose, Costa Rica, Vol. II, pp. 941.

Gunn, R. H. (1955) : The use of Aerial photography in soil survey and mapping in the Sudan, *Soils and Fertilizers.*, 18, pp. 104-06.

Haan, J.E. de and J.A.E. Hendriks (1954) : Lely drop project, a pilot scheme for land development in Surinam, Dutch Guyana, *Journal Agri. Science.* pp. 130-33.

Haggett, P. (1972). *Geography a modern synthesis,* Harper & Row, N.Y. Remote sensing as a Gateway of Environmental Sciences. In Barret, E. C. and Curtis , L . F . (1976). *Introduction to Environmental Remote Sensing.*

Hampton, D. (1961) : *Statistical Analysis of Soil Sampling Ph. D. Thesis,* Perdue University, West Lafayetta, Indiana, U.S.A.

Harris, S.A. (1959) : The classification of Gilgayed soils, *Journal Soil Science* 10, pp. 25-33.

Harvey, D.W. (1968) : Processes pattern and scale problem in geographical research. *Transaction of the Institute of British Geographers,* Vol. 45, , pp. 71-78.

Hasselo, H. N. (1961) : *The Soils of the Lower Eastern Slopes of the Cameroon Mountain,* Wageningen.

Heath, G . R. (1956) : A comparison of the two basic theories of land classification and their adaptability to regional photo-interpretation key technique. *Photogrammetric Engineering* 22, pp. 144-168.

Heath, G.R. (1957) : Correlations between man's activity and his environment, *Photogrammetric Eng.* 23, 108-14.

Hempenius, S.A. (1962) : Specification for mirror stereoscope, *Arch. Int. Photogr.* 14, pp. 45-51.

Hempenius, S.A. (1961) : Dodging a nonlinear element in the contrast transfer of the photographic system, *Photogr.* 16-36.

Heijnsdij, K. D. (1953) : Forestry Photo-interpretation (Dn) Publ. 13, Central Bureau of Aerial Survey, Paramribo.

Holkenbrink, P.E. (1978) : Manual on characteristics of Landsat computer Compatible Tapes Produced by EROS Data Centre Digital Image Processing System USD of Interior, Geologic Survey, Washington D.C.

Imperial Gazetteer of India, Vol. VII.

Indian Geographical Journal (1970) : Vol . XIV, 1970, Madras-5.

Indian Remote Sensing Satellite (1989) : *Data handbook* , Department of space, pp. 4-27.

Irrigation Atlas of India of NATMO (1982) .

Institute Geographi National : *Pondichery (1956) Vegetation maps* 1 : 1M Published Peter Legris'of the Inst.

International Training Centre (ITC) for Aerial Survey (1961) : Prospecters of Department B . Natural Resources Inventor using aerial photographs Delft. I.T.C. Chapter B IX pp. 2-5 (figures).

Jain, A.K. (1971) : *Himalayan Geology,* Vol. I, 1971, pp. 25-58.

Jain, A. K. , Banerjee, D. M. S., Mithal, R. (1971) *Himalayan Geology,* Vol. II. p. 7.

Jarvis, R. (1962) . The use of photo-interpretation for detailed soil mapping Arch., *Inst. Photogrammetry,* 14, 177-82.

Jenkins, D.S., D.J. Belcher, L.E. Gregg and K.B. Woods (1946) : The origin, distribution and airphoto identification of U. S. soils, civil Aeronautics Admn., Techn. D Report No. 52, Washington D.C.

Jenny, H. (1941) : *Factors of Soil formation,* McGraw Hill, New York.

Jones, R. G. B. (1958) . The application of airphoto analysis to soil survey and land classification, *Rhodesia Agri.* J.2. , 195-201.

Jones, C.F. (1959). Status and trends of Geography in the United States. A report prepared for Commission on Geography. Pan American Inst. of Geography and History. *The Professional Geographer* II. part 2, 157 pp. (Jan.).

Junchao and others (1980) : Geologic application of remote sensing in Tengchong, Yunnan China *Proceedings of the XIV Symposium on Remote Sensing of Environment*, San Jose, Costa Rica Vol. III, p p . 1891-1900 .

Kellong, Ch. E. (1951) : Soil and land classification Journal, *Farm Economics* 33, pp. 494-513.

Kellong, Ch. E. (1954) : Soil conservation, Trans. V. Int. Soil Science Congress, *Leopoldville,* Vol. I, pp. 83-102.

Kellong, Ch. E. (1955) : Soil Survey in modern farming. *Journal Soil Water Conservation* - 10, pp. 271-77.

Kellong, Ch. E. (1961) : Soil interpretation in the soil survey, U.S. soil conservation service, Washington, D.C.

Keighley, J . R. , W . W . Lynn and K . R. Nelson (1980) : U se of landsat images in the exploration Brazil, IVX Symp. International on Remote Sensing of Environment, *San Jose, Costa Rica,* Vol. I, pp. 341-43.

Kohn, C. F. (1951) : The use of aerial photographs in the geographic analysis of rural settlements. *Photogrammetric Eng.* 17. pp, 759-71.

Kohn, C.F. *et al.* (1954) : *Settlement Geography, American Geography, Inventory and Process,* pp. 124-41, Syracause University Press. Kral, J. (1952). *Textbook of Aerial Geography and the Geographical Utilization of Aerial Photographs,* pp. 29-31, XVII IGC, Washington.

Kotani, A. (1954) : *Air-photographic Study of Vegetation Coverage of Ozeghara Moor* (Japan).

Leggett, R.E. Ed. (1961) : *Soils in Canada Royal Society,* Canada Space Publ. No. 3, Toronto Press.

Leneuf, J. D. (1960) : Soil science research and agricultural planning in forest and savana (Ivory Coast), U.N. Nat. Resource Conference, Geneva, C 3, 1.

Levenhangst, A.I. (1930) : The use of aerial photographs in Soil Survey Pochvovedenie 4, pp. 116-22 (Russian) , translated. Lindsay, J. D. and Odynsky, Wm. (1956) . Exploratory soil survey by helicopter, *Agri. Inst. Review,* 11, pp. 11-12.

Lewis, A. B. (1952) : Land classification for agricultural development FAO Dev. paper 18, Agri. Rome.

Lowman, P.P. Jr., J.A. McDivitt and E.H. White II (1967) : Terrain Photography on gemmi IV Mission Preliminary Report; NASA, Technical note D-3982.

Lueder, D. R. (1959) : Airphoto interpretation as an aid in mineral reconnaissance and development, *Photogr. Eng.* 19, pp. 819-30.

Lueder, D. R. (1959) : *Principles and Application of Aerial Photographic Interpretation,* Mc. Graw Hill, New York, London.

Lwerovsky, Yu. A. (1958) : Application of aerial methods in the study of soils (Russian) translated, *Pochyvovedenie,* 6, pp. 1-10.

Lwerovsky, Yu. A. and Simahova (1958) : Use of Aerial Photography in Soil Science, I Deleg. *Congress, Soviet Sci. Akad,* Nank, USSR.

Macdonald, D. E. (1953) : Interpretability, *Photogrammetric Eng.* 19, p p . 102-07 .

Macdonald, D. E. (1958) : Resolution as a measure of interpretability, *Photogr. Eng.* 24, pp. 58-62.

Marbut, C.F. (1935) : Soils of U.S. Atlas of American Agri. Part III, U.S. Dept. of Agri. , Washington.

Marchesini, C.A. Pistolesi and A. Bolignim (1962) : Fracture patterns of the natural stream area Larderello (Italy) from airphotographs, *Arch. Int. Photogr.* 14, pp. 524-532.

Marius, C.J. Pias and P. Pisot (1963) : The use of aerial photographs in preparing the pedological map of the Chad Territory on a scale 1:20,0000, U.N. Nat. Resource Conf. Geneva C. 3.1.

Markham, B.J. Townshend and M. Labovitz (1981) : Land cover classification Accuracy as a Function of Sensory spatial Resolution, NASA, T. M. 82071, Goddard Space Flight Center, Greenbelt, Maryland, pp. 8. 14-18. 16.

Maryasu, T. and Nishio (1962) : Experimental studies on colour Air-photographs in Japan. *Photogrammetria* 18, pp. 87-106.

Matsui, T. Ed. (1961) : *Pedologist. Res. Inst. f. Nat. Resource,* Tokyo, pp. 122.

Mayer, H.M. Ullman, E.L. (1954) : Urban Geography. In : *American Geography inventory and prospect,* pp. 142-166. Syracause University Press.

Mc Beth, F. H. (1961) : Aerial photographic investigation of teaching and sopping as an erosion process. *Photogr. Eng.* 27, pp. 154-55.

Mc Fadden (1952) : The uses of aerial photographs in geographic research. *Photogrammetric Eng.* 18, pp. 732-37.

Mc Mohan, C.A. [1877) . The Plain Group and Central Gneiss in Simla Himalaya Rec. *Geol. Survey of India,* Vol. 3, Part-2.

Meijerink, A. M . J. (1982-83) : *Dynamic Geomorphology of Mahanadi Delta ITC Journal,* Special Verstappen Vol. 243-50.

Mikhaulov, V.Y. (1961) : The use of colour sensitive films in aerial photography in USSR. *Photogrammetria* 17, pp. 99-104.

Miles, R. D. (1952) : Application of Aerial Photographs to preliminary engineering soil surveys, *Perdue Eng. Rep.* No. 77, Indiana, USA.

Miles, R. D. (1953) : The use of aerial photos in engineering planning, *Perdue Eng. Repr.* No. 91.

Miles, R. D. (1962) : A concept of land form, parent materials and soils in airphoto interpretation studies for engineering purposes, *Ach. Int. Ph.* 14, pp. 462-76.

Millar, C.E. (1932) : The use of aerial photographs in the Michigan land economic survey, *Amer. Soil Survey Bult.* 13, pp. 82-86.

Miller, R.G. (1960) : The interpretation of tropical vegetation and crops on aerial photographs, *Photogrammetria* 16, pp. 232-41.

Miller, V. G. (1953) : Some factors causing vertical exaggeration and slope distoration on air photos. *Photogr. Eng.* 19, pp. 592-607.

Mitchell, J. (1950) : Productivity ratings and their importance in the soil survey report. Tr. 4th ICSS. Amst. I, pp. 356-66.

Morain, S.A. (1978) : Transferring remote sensing technology . Activities at T.A.C. *Proceeding of the Sensing of Environment*, p p . 1199-1205 .

Mott, P.G. (1956) : The contouring of a tropical forest area in Ceylon by the use of aerial photographs, *Emp. For. Review,* 35, pp. 36-41.

Muir, A. (1955) : The use of airphotographs in Soil Survey, *Photogr.* 6. pp. 50-57.

Muralikrishna, I.V. (1992) : *Remote Sensing Application and Geographic Information System.*

Nag, P. and R. K. Banerjee (1992) : *Thematic Cartography and Remote Sensing.*

Nakano, T. (1962) : Landform type analysis on aerial photographs, its principle and technique, *Arch. Int. Photogr.* 14, pp. 149-52.

Nakano, T. , Tahasaki; M. and Kawai, R. (1962) . Geographical surveys by means of aerial photo-interpretation , in Japan, *Arch. Int. Photogr.* 14, pp. 307-10.

Nash, A. J. (1958) : Template for assisting photo-interpretation, *Journal of Forestry*, Vol. 56, Nu. 8, p. 594.

National Geographical Journal of India (1955) : vol . I, Part I.

Niyogi, D. (1975) : Geology of Coastal Plain in West Bengal and Orissa. *Indian Journal of Earth Science* 2 pp. 51-61.

Odell, R. T. (1958) : Soil Survey interpretation yield prediction. *Soil Science Society American Proceeding.* 22, pp. 157-60.

Oldham, R.D. (1883) : Note on Geology of Jaunsar and Lower Himalaya, *Geological Survey of India,* Vol. 16, pt. 4, page 193-98.

Orfanitsky, Yn. A. (1952) . The relationship between soils and relief and the use of aerial photographs in detailed soil mapping. *Pochvovedenie,* 332-40, Moskov (Russian) translated.

Paul, K. and Mascarenhas, C. (1981) : *Remote .Sensing in Development Science,* Vol. 214, no. 4517, pp. 139-45.

Pachman, D.J. and L. E. Philpotts (1955) : *Elementary Agricultural Airphoto-interpretation,* Dept. Agri. Ottawa.

Pasto, J.K. (1953) : Soil mapping by Stereographic interpretation of airphotos. *Soil, Sci. Socy. American Proceed* 17, pp. 135-38.

Peltier, L.C. (1954) : *Geomorphology, American Geography, Inventory and Prospect.* pp. 362-81. Syracause Univ. Press.

Pena Rodrigues, F. (1963) : Soil Study in Mexico by interpretation of aerial photographs, U. N. Nat. Res. Conf. Geneva, C 3.1.

Perrin, F. H. (1966) : *Structure of the Developed Image,* Mc Millan, New York, pp. 499-551.

Perrin, R.M.S. (1962) : The use of airphotographs in the study of patterned ground in East Anglia, *Res. Int. Photogramm.* 14, pp. 183-88.

Pias, J., C. Marins, and P. Poisot (1963) : Use of serial photographs in preparing the pedological map of the Chad Territory on a scale 1:2000,000. U.N. Conference, Geneva, C 3.1.

Pilgrim, G. E. and West, W. D. (1928) : *Structure and Correlation of Simla rocks,* Memoirs G.S.I., Vol. 53, p. 140.

Pommeroning, J.A. and M.G. Cline (1953) . The accuracy of soil maps prepared by various methods that use serial photographinterpretation, *Photogr. Eng.* 19, pp. 809-17.

Popov, I. V. (1961) : *Aerial Survey and Study of Land Waters* (Russian), translated.

Population and Landuse map of National Atlas of India (1982). Nagpur plate.

Powers, W.L. (1954) : Soil and landuse capabilities in Iraq. *The Geogr. Rev.* 44, pp. 373-80.

Prapinmong Kolkarn, P.C. Thisayakorn, N. Kaliyakulwanich and others (1980). Remote Sensing Application to land cover classification in Northern Thailand. *Proceeding of the 14th International Symposium on Remote Sensing on Environment,* Vol. II, pp. 1273-84.

Prescott, J.A. and J.K. Taylor (1930) : The value of aerial photography in relation to soil surveys and classification, *Journ. CSIR,* Australian, pp. 229-30.

Quam, L. O. (1952) : The application of photo-interpretation to geographic research. *Photogrammetric Eng.* 18, 500-01.

Raisz, E. (1951) . The use of air-photos for landform maps, *Annals of the Association of American Geographers* 41, pp. 314-30.

Raychoudhuri, S. and B.R. Datta Biswas (1959) : Saline and Alkali soils of Asia with particular reference to India, Vol. I, pp. 191-207, Vth In. Cong. Soil Science, General Lecture.

Retallac, G.J. 1990 : *Soils of the Past,* Unwin Hyman.

Roscoe, J. H. (1952) : *Photo-interpretation keys,* I. G. U., pp. 142-43.

Roscoe, J. H. (1952) : Photo reading, photo-interpretation and photo intelligence with special reference to photogeography, *Proceeding 8th General Assembly and l7th I. C.*, IGU, pp. 144-45. Washington.

Roscoe, J . H. (1953) : Photogeography, selected papers on photogeology and photo-interpretation, R & D, Board Publication, No. G. G. 209/1, pp. 55-102, Washington.

Roscoe, J.H. (1954) : *Photogeography, Photogeology and Photo-interpretation Key,* Berkeley, California.

Roy, B.K. (1992) : *Statistical Thematic Mapping*, pp. 51-68 (Th. C. and R.S.).

Sader, A. (1980) : Renewable resource analysis and monitoring in Costa Rica; *Proceeding of the l4th Int. Symp. on Remote Sensing of Environment, San Jose, Costa Rica,* Vol. 2, pp. 1251-60.

Sadowski, F. and J. Sarno (1976) : *Forest Classification Accuracy as Influenced by Multispectral Scanner Spatial Resolution,* Michigan, p. 130.

Sambalpur District Gazetteer (1971) : Printed by the Spdt. Orissa Govt. Press, Cuttack, Senapati and Oths.

Sambasiva, Rao and others (1978) : Morphology and Evolution of Mahanadi and Brahmani-Baitarani Delta, Dept. of Geology, Delhi University.

Schermerhorn, W . (1960). Planning in modern aerial Survey, *Photogramm,* 17, pp. 1-17.

Sharma, R.P. (1972) . Elucidation of structure and stratigraphy of belt north of the Tons river in H.P. Himalaya, Appreciation seminar in Airphoto interpretation .

Siefferman, G. (1980) : Contribution to soil mapping of Southern Venezuela using computer, analysis of Landsat II data. *Proceeding of the IVXth Symposium on Remote Sensing of Environment.* San Jose, Costa Rica, pp, 1221-22.

Sinha, B. N. (1968) : *India Regional Study : Mahanadi Delta.*

Smith, H. T. U. (1941) : Aerial photographs in geomorphic Studies, *Journal of Geomorphology* 4, pp. 171-205.

Smith, H. T. U. (1951) : Symposium : Airphotos in Geography and Soil Science, *Photogrametric Engineering* 17, pp. 715-79.

Soil Survey Staff (1951) : *Soil Survey Mannual.* IARI.

Spate, O. H. K. (1961) : *Geography of India and Pakistan.*

Srikanta, S.V. and Sharma, R.P. (1969) . Shali formation, *Bull. G.S. of India,* Vol. 6, No. e, p. 93-97.

Srikanta, S.V. and Sharma, R.P. (1969) . Shali formation, *Bull. G.S. of India,* Vol. 6, No. e, p. 93-97.

Stamp, L.D. (1940) . Fertility, productivity and classification of land in Mritain, *Geographical Journal* 96, pp. 380-412.

Steiner, D. (1961) : *Die Jahreszeit ale Factor bei der landesnutzungs-interpretation.*

Stone, K. H. (1953) : World air-photo coverage for geographic research, *Annals of the Association of American Geographers* 43 : 193.

Stone, K. H. (1956) : Airphotos interpretation Procedure, *Photogr. Eng.* 22, 123-32.

Suprato Hardjo, M. (1954) . Utility and practical possibility of aerial photography for cartography of Indonesian soils. *Technik Pertanian*, pp. 109-48 (Indonesia), translated.

Swanson, C.L.W. (1955) . Aerial photography requirements for soil survey field operation, World crops 7, pp. 193-194, also *photogr. Eng.* 1954 (20), 709-11.

Swain, P . H. and S. M. Davis (1978) : *Remote Sensing Quantitative Approach.* Mc Graw Hill, New York.

Taylor, N.H. and I.J. Pohlen (1962) : Soil survey method, N. Z. *Soil Bureau Bulletin* 25.

Theurer, C. (1959) : *Color and Infrared Experimental Photography for Coastal Mapping Eng. 25,* pp. 565-69.

Thurell, J. (1953) : Vertical Exaggeration in Stereomodel, *Photogr. Eng.* 19, pp. 579-92.

Townshend, I.R. (1980) : The Spatial Resolving Power of Earth Resource Satellites. A Review NASA Goddar Space Flight Center, Greenbelt, Maryland, U.S.A. p. 36.

Transport and Tourist Map of India (Calcutta and Nagpur plate 1982) : Prepared by NATMO, Dept. of Science and Technology, Govt. of India, Calcutta.

Troll, C. (1962) : Geografische Luftbildinterpretation, *Arch. Int. Photogrammetia* 14, pp. 266-75.

Tucker, C. J. (1979) : Radiometric Resolution for Monitoring Vegetation — How many Bits are needed? NASA. T.M. 80293, G.S.F. Center, Greenbelt, Maryland, U.S.A., p. 25.

Van der Eyk, J.J. (1957) : Reconnaissance Soil Survey in Northern Surinam (Dutch Guyana).

Van der, K. J. J. and H. A. J. Hendriks (1953) : Soil and land classification in the old coastal plain of Surinam (Dutch Guyana), *Neth. Journal Agri. Science,* pp. 278-98.

Van der Weele, A. J. (1963) : *Photogrammetric en civiele Techniek.* Rede Delft.

Vaidyanathan, R. W. (1991) : *Quarternary Deltas of India,* Geological Society of India, Bangalore.

Veenenbos, J. S. (1956) : (1) Aerial photo-interpretation and analysis for soil survey and land classification purpose.

Veenenbos, J. S. (1956) : (2) Small scale pedological analyses and map compilation VI. *International Soil Science Cong.* Paris, Vol. E, p. 251-25.

Vink, A.P.A. (1961, 1962, 1963) : (a) The aerial photographs in relation to the agricultural development of tropical and sub-tropical region (Dutch, English Summary).

(b) Practical soil survey and their interpretation for practical purpose, ITC Publ. B. 15, Delft.

(c) Planning of soil survey in land development International Institute of land reclamation and drainage, pub. 10, Wageningen.

Wagner, W. and S. Lowe (1978) : A small grant programme for remote sensing technology transfer, *Proceeding of the 12th Int. Symp. on Remote Sensing of Environment,* Vol. 1, pp. 895-906.

Walker, F. (1953) : *Geography from air.* pp. 111. New York, Button.

Weatherhead, T. D. (1951) : Resource Surveys of under development areas within the British Commonwealth and Empire. The application of aerial methods, *J. Royal Soc. Arts* 99, pp. 848-64.

West, W.D. (1939) : *The structure of Shali window near Simla Recd.* , Vol. 74, pt. 1, pp. 113, 163.

Winkler, E.M. (1960) : Interpretation of glacial drift, *Photogr. Eng.* 26, 773-74.

Witen Stein, M.M. (1955) : Uses and limitation of aerial photography in urban analysis and planning. *Photogrammetric Eng.* 21, pp. 566-72.

INDEX